Biotech's

DICTIONARY *Of* Ecology

Compiled & Edited by

Dinesh Arora

BIOTECH

2014

BIOTECH BOOKS

DELHI - 110035

Biotech's

DICTIONARY

Of

Ecology

Compiled & Edited by
Dinesh Arora

ISBN 81-7622-133-3

Published by : **BIOTECH BOOKS**
1123/74, Tri Nagar,
Delhi - 110 035
Phone: 27382765
e-mail: biotechbooks@yahoo.co.in

Showroom : 4762-63/23, Ansari Road, Darya Ganj,
New Delhi - 110 002
Phone: 23245578, 23244987

Printed at : **Chawla Offset Printers**
New Delhi - 110 052

PRINTED IN INDIA

PREFACE

'Ecology' as understood is the study of livings and deeds in their ecosystems. It is a huge subject that includes many of the rules and laws set up within physics, chemistry, geology, oceanography, physiology and biology. Ecosystem is the relationship between living and non-living things in different environments. As a result, ecology depends on how organisms, whether humans, animals or plants, coexist and interact with each other.

The science of ecology is concerned with relationships between living organisms: plants, animals, microorganisms and their environment. Ecologists study the way in which organisms, populations, communities and ecosystems function, and in so doing also draw on many other areas of biology including genetics, evolution, physiology and behaviour. Ecology is thus a challenging and diverse subject. Here, a wide range of environmental disciplines, ranging in scale from the gene to whole Earth systems have been specially highlighted. The aim is to improve our understanding of both the environment and the processes that underlie the Earth's support systems. Moreover, the impact of human activity on natural environments are particularly stressed. Therefore, it is necessary to generate practicable solutions to today's pressing environmental problems, so that a healthy, wealthy and sustainable environment can be enhanced and maintained.

Vibrant neighbourhoods are the building blocks of healthy

cities and thriving regions. Ecology uses urban as well as rural design, land use planning and policy reform to help people plan and build communities that are ecologically healthy, socially just and economically fair.

- **a**
  in population genetics, a symbol designating a particular allele.

- **abaptation**
  1. the process by which the present match between organisms and their environment and the constraints on this match, have been determined by evolutionary forces acting on ancestors.
  2. the prefix 'ab-' emphasises that the heritable characteristics of an organism are consequences of the past and not anticipation of the present or future.

- **abiotic**
  1. usually applied to the physical and chemical aspects of an organism's environment.
  2. non-living. Climate is an abiotic component of ecosystems.
  3. a term pertaining to substances or objects that are non-biologic in origin.

- **abiotic factors**
  environmental influences produced other than by living organisms; for example, temperature, wind patterns, humidity, pH, substrate rock type and other physical and chemical influences.

- **abnormal behaviour**
  1. may occur as a result of any pathological condition in an animal.
  2. the term is often associated with experimentally induced or captivity related anxiety.

- **absolute poverty**
  the lack of sufficient income in cash or exchange items for meeting the basic needs of food, clothing and shelter.

- **abundance**
  the number of organisms in a population combining 'intensity' (density within inhabited areas) and 'prevalence' (number and size of inhabited areas.)

- **acceleration**
  a speeding up of development in ontogeny so that a feature appears earlier in the ontogeny of a descendant than it did in the ancestor.

- **acceptable limits of change**
  1. the limits to the type and scale of change appropriate to an area.
  2. The limits may be due to environmental, social or

economic concerns. What is acceptable or appropriate is determined by consultation with governments and communities, as well as by legislation and regulations.
3. The limits of acceptable change establish the maximum 'damage' level for a resource that society is prepared to accept.

- **acclimation**
  the habituation of an organism's physiological response to environmental conditions, usually applied to laboratory environments.

- **acclimatisation**
  1. the habituation of an organism's physiological response to environmental conditions, usually applied to natural environments.
  2. a form of reversible physiological adaptation by which an animal is able to alter its tolerance of environmental factors.

- **acid fallout**
  molecules of acid formed from reactions high in the atmosphere involving nitrogen, sulphur oxides and water vapour that settle out of the atmosphere without any additional water.

- **acid precipitation**
  includes acid rain, acid fog, acid snow and any other form of precipitation that is more acidic than normal (i.e., less that pH 5.6). Excess acidity is derived from certain air pollutants, namely sulphur dioxide and oxides of nitrogen. The effects can include: death of fish, and *eutrophication* of lakes; damage to tree, leading to soil erosion and physical corrosive damage to vehicles and buildings.

- **acid rain**
  rain with a very low pH (often below 4.0) resulting from emissions to the atmosphere of oxides of sulphur and nitrogen.

- **acid sulphate soils**
  soils rich in iron sulphide common in coastal areas. They are relatively stable while saturated with water,

but the sulphide forms sulphate and sulphuric acid when drained and exposed to air. The sulphuric acid can kill fish, prawns, oysters and other aquatic life when washed into waterways by rain. It can also corrode concrete slabs, steel fence posts, foundations of buildings, and underground concrete water and sewerage pipes and weaken concrete structures. It can be very difficult to revegetate iron sulphate soils once they become exposed.

- **acoustic**
referring to hearing and sounds.

- **acquired behaviour**
behaviour that has been experimentally demonstrated, in either its topography or stimulus control or both.

- **Acquired Releasing Mechanism**
the ARM is a releasing mechanism which an individual has acquired during its ontogeny.

- **acquisition**
1. progressive increments in response-strength observed over the series of occasions on which the response is measured.
2. any modification of behaviour in which a response changes in strength or topography, or occurs in new environments.

- **action pattern**
a series of actions having consistent form in a species. See **fixed modal action pattern**.

- **action potential**
1. like, transient, all-or-none reversal of membrane potential that sweeps along the membrane of a nerve or muscle cell. When stimulated the cell membrane changes conductance for various ions. Sodium ions rush in and the electrical charge across the membrane becomes positive. Local membrane depolarisation stimulates the neighbouring region to depolarised and then returns to its resting state.
2. changes in electric potential on the membranes of living cells, resulting from cell stimulation, leading to an all-or-nothing action current. An abrupt change

in the charge differential across the membrane of a nerve cell caused by a change in the permeability of the membrane to sodium ions.

- **action-specific energy or Specific Action Potential (SAP)**
  the energy that accumulates. This is a hypothetical construct inferred from changes in stimulus threshold, intensity and rate of occurrence of an unlearned response over time and with the frequency of occurrence of the response. SAP decreases when a response is performed; it then recovers with time in the absence of a response.

- **activational effect**
  the capacity of key hormones to cause certain behavioural responses to occur by activating proximate mechanisms underlying the behaviour.

- **active transport**
  movement of ions or molecules against an electrochemical gradient by utilisation of metabolic energy.

- **activity**
  a crude behavioural category to measure overall movement of the animal.

- **ad lib sampling**
  a method of sampling behaviour where animals are chosen opportunistically.

- **ad libitum**
  at pleasure, meaning unrestricted access to food or water.

- **adaptation**
  1. a confusing word used to mean quite different things.
  2. characteristics of organisms evolved as a consequence of natural selection in its evolutionary past and which result in a close match with features of the environment and/or constrain the organism to life in a narrow range of environments.
  See **Abaptation**.
  3. changes in the form or behaviour of an organism *during its life* as a response to environmental stimuli, e.g. the formation of sun and shade leaves on the same tree and the acquisition of cold tolerance as a result of prior experience of low temperatures.

4. changes in the excitability of a sense organ as a result of continuous stimulation.
5. decline in the amplitude of the response of a sensory neuron in the presence of a constant stimulus.
6. a process of genetic change of a population, owing to natural selection, whereby the average state of a character becomes improved with reference to a specific function, or whereby a population is thought to have become better suited to some feature of its environment. Adaptation is a feature that has become prevalent in a population because of a selective advantage owing to its provision of an improvement in some function.

- **adaptationist**

one who believes that most characters are due to the action of natural selection; called 'aptation' by behavioural biologists who uses the assumption that traits are adaptive to generate testable hypotheses on the fitness consequences of a particular trait.

- **adaptive landscape**

The notion of adaptive landscapes refers to the fact that there are many such local optima and that animals cannot easily evolve from one to another because of the drop in fitness that is associated with the intervening phenotypes. See **adaptive peak**.

- **adaptive management**

1. a type of natural resource management implies making decisions as part of an on-going process. Monitoring the results of actions will provide a flow of information that may indicate the need to change a course of action.
Scientific findings and the needs of society may also indicate the need to adapt resource management to new information.

- **adaptive peak**

1. refers to the notion that combination of alies at two or more loci are adapted to a specific function or to some feature of the environment such that any deviation from that phenotype results in a reduction in fitness.

2. genotypes or phenotypes associated with suites of morphological characters that have high fitness with the areas between being short-lived evolutionary phenomena.
3. That allele frequency, or combination of allele frequencies at two or more loci, at which the mean fitness of a population has a (local) maximum.

- **adaptive radiation**
1. the evolution of a single ancestor species into several new species within a relatively short period of time and in a certain geographic area. The plants and animals of the Galapagos Islands are a result of adaptive radiation, where one plant or one animal species diversified into many species that fill a variety of ecological roles. For example, more than a dozen species of finches evolved from a single founding species that colonised the islands from the mainland of South America.
2. evolutionary divergence by the members of a single phylogenetic line into a number of distinctive types or adaptive forms.

- **adaptive significance**
the function or survival value of a behaviour.

- **adaptive value**
the fitness of a trait.

- **adaptive zone**
a set of similar ecological niches occupied by a group of (usually) related species, often constituting a higher taxon.

- **additive effect**
of an allele, its effect on the phenotype, averaged over the variety of genetic combinations in which it occurs. It is said to act additively if its effect is independent of the other alleles.

- **Adenosine Tri-Phosphate (ATP)**
molecule composed of adenine, ribose and three phosphate groups bound by high-energy linkages and associated with energy transfer in living cells.

- **adiabatic expansion**
expansion taking place without heat entering or leaving the system.

- **adipsia**
  the lack of thirst.

- **adjunctive behaviour**
  behaviour induced by an intermittent reinforcement schedule characteristically displayed during the time interval when reinforcement is highly improbable.

- **adoption**
  individuals care for young to whom they are not related.

- **adventitious reward**
  a reward that is unplanned by the experimenter or not under the experimenter's control.

- **adventitious roots**
  roots that arise in 'abnormal' positions, e.g. from a stem or leaf. The contrast is with the primary roots that develop from the axis of a seedling and roots that arise from other roots.

- **advergent evolution**
  when the ancestor of a model has warning coloration and the ancestor of a mimic evolved toward the model so that it came to look like the model.

- **advertisement**
  a form of conspicuous display. Territorial advertisement refers to behaviour patterns that serve to attract females and ward off conspecific intruders.

- **aerial logging**
  removing logs from a timber harvest area by helicopter. Fewer roads are required, so the impact to an area is minimised.

- **aerobe**
  an organism that utilises atmospheric oxygen in its metabolic pathways. An organism that must have oxygen in order to survive is an *obligate aerobe*.

- **aerobic decomposition**
  the process of breakdown of organic molecules to simple inorganic constituents when oxygen is in free supply.

- **aerobic**
  1. living or occurring only in the presence of oxygen: aerobic bacteria.
  2. of or relating to aerobes, organisms that require and utilise oxygen.

3. involving or improving oxygen consumption by the body: aerobic exercise.

- **aerophagia**
  the swallowing of air.

- **aesthetic injury level**
  the level of pest abundance above which aesthetic or sociological considerations suggest control measures should be taken against the pest.

- **aestivation**
  1. a state of dormancy during the summer or dry season.
  2. a period of dormancy during the summer in contrast to hibernation which is winter dormancy.

- **affected environment**
  the natural environment that exists at the present time in an area being analysed.

- **afference**
  excitation, impulse or information moving from the periphery over afferent nerve fibres into the central nervous system in contrast to efference.

- **afferent inhibition**
  inhibition of the IRM resulting in the termination of an animal's response to specific stimuli. However, if the stimuli are changed somewhat, then the animal may again respond.

- **after-discharge**
  1. that part of a response that occurs after the termination of the stimulus that elicited it or set the occasion for its occurrence.
  2. the term applied to discharges of nerve impulses in efferent neurons that persist in time after the stimulus that set up the discharge is terminated.

- **age class**
  an age grouping of trees according to an interval of years, usually 20 years. A single age class would have trees that are within 20 years of the same age, such as 1-20 years or 21-40 years.

- **aggregated distribution**
  the distribution of organisms in which individuals are closer together than they would be if they were randomly or evenly distributed.

- **aggregation of risk**
  a pattern in which prey vary, from prey patch to prey

patch, more than would be expected by chance alone in their risk of being attacked by a predator (especially applied to hosts varying in their risk of being attacked by parasitoids.)

- **aggregation**
a group which is formed by a mutual attraction to a common resource such as a feeding aggregation (in contrast with social group.)

- **aggregative response**
the response of a predator through which it spends more time either in habitat patches with higher densities of prey, leading to higher densities of predators in patches with higher densities of prey, or in habitat patches with lower densities of prey, leading to higher densities of predators in patches with lower densities of prey. See also **Aggregation of risk**.

- **aggression**
a physical act or threat of action by one individual that reduces the freedom or genetic fitness of another. Aggression may refer either to the behaviour or to the underlying motivation. There are several different kinds of aggression that almost certainly involve different sorts of controlling mechanisms.
1. fear-induced aggression occurs only in cases where escape has been attempted but it not possible.
2. predatory aggression is elicited during predation.
3. male-male or inter-male aggression is specific to adult breeding males and is generally associated with guarding females.
4. irritable aggression refers to aggression resulting from frustration, deprivation or fear.
5. territorial aggression appears in defence of a territory.
6. maternal aggression appears in defence of young.
7. instrumental aggression consists of an increase in the tendency for an organism to engage in aggressive behaviour when that behaviour has been reinforced in the past.

- **aggressive behaviour**
a broad class of behaviour that includes both threat and attack behaviour.

- **aggressive mimicry**
  mimicry in which a predator mimics a non-predatory model in order to deceive the operator, its prey.

- **agonistic behaviour**
  refers to the behaviour associated with aggression and escape behaviour. A broad class of behaviour that includes all attack, threat, appeasement and flight behaviour.

- **agonistic buffering**
  the use of infants by adults to inhibit the aggressive behaviour of other adults.

- **agroforestry**
  production of tree crops in a manner similar to agriculture. Also, production of trees along with regular crops.

- **aha-erlebnis**
  the sudden insightful understanding of a previously unknown relationship. This sudden insight which is not accomplished in a stepwise manner is often associated with the exclamation, 'Ah – Ha'.

- **airshed**
  a geographic area that shares the same air.

- **akinesis**
  immobility determined by reflex torpidity resulting from continued contraction of the locomotory muscles, for example, freezing of some animals when in danger.

- **alarm call**
  a sound that warns others of the presence of predators.

- **alarm pheromone**
  the alarm substance which arouses other members of the social group to defend the nest from enemies or invaders.

- **alcohol dehydrogenase**
  an enzyme catalysing the conversion of alcohols into aldehydes and ketones and the reverse.

- **Alfred Russel Wallace**
  a contemporary of Darwin who also is credited with helping to formulate the concept of natural selection as a mechanism of evolution.

- **allate**
  winged.

- **allele**
  any of the two or more alternative forms of a gene

that occur at the same locus on a chromosome.

- **allelic frequency or gene frequency**
  the frequency of a given allele as a proportion of all alleles at that locus in a population.

- **allelochemical**
  a substance produced by one organism that is toxic or inhibitory to the growth of another.

- **allelomimetic behaviour**
  behavioural activities that have strong components of social facilitation, imitation and group coordination.

- **allen's rule**
  the principle that states that in warm-blooded species, the relative size of exposed portions of the body (limbs, tail, ears) decreases with a decrease of mean temperature.

- **allochronic**
  sympatric species or populations that are reproductively isolated due to mating at different times.

- **allochthonous behaviour**
  1. behaviour that is not activated by its own drive.
  2. behaviour that is activated as a consequence of frustration, or behaviour that is activated by some drive other than that which most often controls it in contrast to autochthonous.

- **allochthonous material**
  organic matter entering a stream, lake or ocean but derived from an adjacent terrestrial system.

- **allochthonous**
  formed or occurring elsewhere than in place; of foreign origin.

- **allofeed**
  to feed young that are not the offspring of the feeder; a special case of alloparenting.

- **allogenic succession**
  a temporal succession of species at a location that is driven by external influences which alter conditions.

- **allogrooming**
  social grooming.

- **ambivalent behaviour**
  behaviour that consists of two different, often incom-

plete reactions occurring simultaneously or in short rapid succession. This occurs in conflict situations.

- **amble**
  in the horse a variation of the pace gait that is distinguished by a slight hesitation between the placing of the hooves of the front and rear legs on the same side of the horse. The amble is a slower gait than the pace.

- **ambush predator**
  an animal that awaits its prey and ambushes it rather than stalking it.

- **amensalism**
  1. an interaction in which one organism or species adversely affects a second organism or species, but the second has no effect on the first.
  2. in a social interaction where care or resources are involved, a loss to the donor but neutral to the recipient.

- **amino acids**
  the nitrogen-bearing molecules that can be polymerised to form proteins; the basic chemical units of proteins. Life as we know it consists of twenty major varieties.

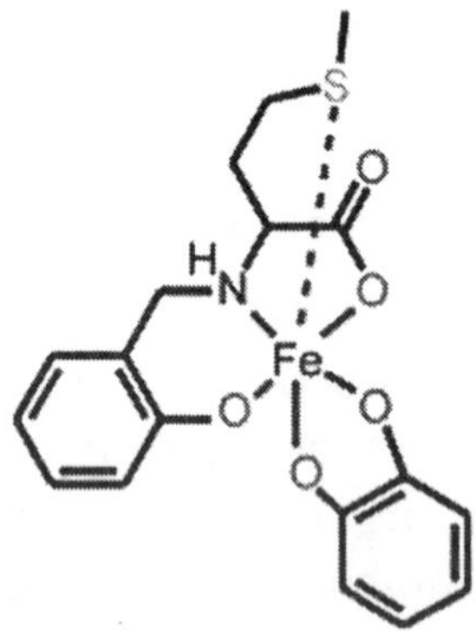

- **amplexus**
  the clasping mating behavi-our of toads or frogs in which the male sits on the female's back as he clasps her sides.

- **amplitude of response**
  a quantitative measure of one dimension of a response. Response magnitude.

- **anadipsia**
  intense thirst.

- **anadromous**
  migration in which adults spend lives at sea and return to freshwater to spawn.

- **anadromous fish**
  species of fish that mature in the sea and migrate into streams to spawn.

- **anaerobe**

1. an organism capable of living in the absence of free oxygen (O_2).
2. obligate anaerobe: An organism that must live without oxygen, for whom oxygen (O_2) is toxic.

- **anaerobic decomposition**

a process of breakdown of organic molecules to simpler inorganic constituents, that occurs in the absence of oxygen.

- **anaerobic**

1. lacking or seriously depletion of oxygen. Opposite of aerobic.
2. of or relating to organisms, such as certain bacteria, that can live in the absence of atmospheric oxygen (indeed, for most anaerobic bacteria, oxygen is toxic).

- **anagenesis**

describes directional evolution of a feature over an arbitrarily short segment of a lineage.

- **analogous structures**

organs of different evolutionary origin which perform the same role in different organisms.

- **analogy**

1. resemblance in characteristics (e.g., body form, behaviour) as a consequence of independent adaptation to the same or similar environmental conditions and not due to common ancestry. Related Terms: **homology**
2. a resemblance in function and often in appearance, between two structures, physiological processes or behavioural patterns that are due to independent evolution rather than common ancestry. See **homology**.

- **analysis of covariance**

a statistical method for determining whether the functional relationships described by two of more regressive equations are the same, i.e. could have come from populations with the same slope; it comprises a combined application of linear regression and analysis of variance and is used when treatments are compared in the presence of accompanying variables which cannot be eliminated or regulated.

• **analysis of variance**
a statistical technique for partitioning the total variability affecting a set of observations between the possible and statistically independent causes of the variability; useful for examination of sources of uncontrolled variation in experimental situations.

• **anaxonal neuron**
neuron that has no neurites or extremely short ones. These neurons are typically found in sensory systems, such as the vertebrate visual and auditory systems.

• **androgen**
generic term for male hormones.

• **anemotaxis**
a transverse orientation behaviour in which animals orient with respect to wind direction.

• **aneuploidy**
of a cell or organism, possessing an unbalanced chromosome complement by virtue of an excess or deficiency in number of one or more of the chromosomes compared to the others.

• **angiosperms**
flowering plants. Strictly, those seed-bearing plants that develop their seeds from ovules *within* a closed cavity, the ovary (*cf.* gymnosperms, in which the ovules are naked.)

• **angst**
an effect or feeling associated with oppression or persisting despair. May be associated with physical symptoms such as defecation, urination, sweating of the palms etc.

• **animal behaviour**
wild and wonderful ways in which animals interact with each other, with members of other species, and with the environment; related term: behaviour.

• **anisogamy**
1. condition in which the female gamete is larger than the male gamete; contrast with Isogamy.
2. refers to condition in which the male and female gametes are of different sizes.

• **ankle attacks**
cats commonly attack the ankles of someone entering

the room or any moving object. The easiest way to stop these attacks is to provide the cat with movable toys.

- **annual**
  a species with a life cycle which takes approximately 12 months or rather less to complete, whose life cycle is therefore directly related to the annual cycle of weather, and whose generations are therefore discrete.

- **anorexia**
  occasional failure to eat is common in animals and may be caused by unpalatable food, nausea or individual variation in appetite. Long-term anorexia is caused by lesions in the hypothalamus or medical problem that act on the GI system or brain.

- **anoxic**
  1. without oxygen.
  2. pertaining to the absence of uncombined molecular oxygen.

- **antagonistic resources**
  a pair of resources for which increased consumption by an organism of one resource leads to an increased requirement by that organism for the other.

- **antennation**
  touching with the antennae.

- **anthropocentric**
  regarding humans as the central fact of the universe and interpreting everything in relation to them.

- **anthropogenic**
  caused or produced through the agency of humans.

- **anthropomorphism**
  refers to the giving of human attributes to animals. Critical anthropomorphism refers to using anthropomorphisms to think through or explain a problem.

- **anticipatory action**
  see **intention movement**.

- **anticodon**
  the triplet of nucleotides in a tRNA molecule which associate by complementary base pairing with the codon in the mRNA during translation.

- **anting**
  an innate behaviour in several bird species that serves to spread the poisonous secretions of ants over the plumage. Some species sit

down directly on an anthill whereas others hold ants in their beach and rub them over their plumage.

- **antiphonal singing**
  a type of duetting in which the birds sing with alternated calls and responses.

- **anti-predator behaviour**
  1. behaviour patterns that have evolved as defence mechanisms against predators.
  2. generalised term for all naturally occurring mechanisms for escaping a predator. This includes protective coloration, as well as a variety of behavioural adaptations such as death feigning, distraction displays, threat and outright defence.

- **aphrodisiac**
  a chemical which when produced makes a female more likely to mate with the sender.

- **apomixis**
  parthenogenetic reproduction in which an individual develops from an unfertilised egg or somatic cell.

- **apomorphic**
  apomorphy. A derived character state based upon comparison with the ancestral state.

- **aposematic behaviour**
  1. behaviour patterns which neutralise or redirect the aggressive attacks of conspecifics and often serve to change the 'mood' of the attacker.
  2. those behaviours of an animal (exclusive of flight) that, when they occur, terminate attack by another animal of the same species.

- **aposematic**
  colouration or other features that advertises noxious properties; warning colouration.

- **aposematism**
  conspicuous appearance of organism that is noxious or distasteful.

- **apostatic selection**
  differential survival of one morph due to the action of predators who forms learned associations between particular visual images and acceptable food.

- **apparent competition**
  an interaction in which the organisms (or species) have

adverse effects on one another by virtue of the beneficial effects that each has on a predatory organism (or species) which they share.

- **appeasement**

behaviour that serves to inhibit or reduce aggression between members of the same species often in situations where escape is impossible or disadvantageous.

- **appetite**

refers to those physiological changes and cognitive events that are evoked by the stimulus properties of foods. Hunger on the other hand refers to the physiological changes and cognitive events which are the consequence of prolonged fasting.

- **appetitive behaviour**

1. specific behaviour in which an individual searches for a releasing stimulus situation which leads to the desired consummatory act by means of the releasing mechanism. Often used loosely to describe the wandering behaviour that often precedes a consummatory act.
2. a term applied to characterise, in terms of an inferred or anticipated (by the ethologist) consummatory act, the behaviour of an animal that is not at rest or doing nothing. Traditionally this refers to the active, goal-seeking and exploratory phase of behaviour that precedes the more stereotyped consummatory behaviour that an animal exhibits when it reaches its goal. Now regarded as a conceptually vague notion that is difficult to define operationally. Appetence for aggression would refer to an animal seeking out a stimulus that would elicit aggressive behaviour.

- **applied ethology**

a branch of behaviour study concerned with animal species that are of direct practical interest to people, emphasizing the possibilities for practical application that are within ethology's domain. As most of the species with which people deal directly are other domestic animals or zoo animals, applied ethology more or less divides between veterinary ethology and zoo biology.

• **approach-withdrawal theory**
according to the theory, when animals begin life they are capable only of forced movements; intense stimulation causes withdrawal, and mild stimulation causes approach. By interacting with its environment, the organism builds up this base of qualitatively distinct levels of behavioural capacity, such as discrimination, motivational control and hypothesis formation.

• **approximation conditioning or shaping conditioning**
if the experimenter wants to condition a highly improbable (infrequent) operant response (one that will not occur and, hence, cannot be reinforced, except rarely such as bar pressing), the experimenter may shorten the tie before the subject emits the response by reinforcing other responses that are successively more like the response to be conditioned.

• **apterous**
wingless.

• **aquifer**
a body of rock that is saturated with water or transmits water. When people drill wells, they tap water contained within an aquifer.

• **arboreal**
1. living predominantly or entirely in trees.
2. living in trees; adapted to life in trees.

• **area effect**
abrupt geographic chang in morph frequencies between two relatively uni form populations or serie of populations living in a apparently uniform habita

• **area effects speciation**
a kind of parapatric speciation in which reproductive isolation develops betweer two populations that show area effects.

• **arena display**
see **lek**

• **arenicolous**
living in sand.

• **arms race**
the notion that animals ar constantly evolving adapta tions to counter the adapta tions of their predators, con

petitors or parasites. Each evolutionary advance by one party is countered by an advance by the other. An arms race is 'won' when one party drives the other extinct.

- **arousal**

refers to several different concepts:

1. the original usage in physiology distinguishing states of sleep and wakefulness,
2. distinguishing levels of responsiveness from periods when the animal is willing to do things to periods when the animal does not respond except with high levels of stimulation,
3. distinguishing levels of activity and the kinds of behaviour being performed (e.g. grooming low level, fighting high level),
4. distinguishing the degree of reflex preparation for exertion or immobility.
5. the amount of sensory input which the central nervous system is prepared to accept at a particular time.

- **artenkreis**

groups of species that replace one another geographically as distinct from Rassen Kreis.

- **arthropod**

a member of the animal phylum Arthropoda, which includes the insects, crustaceans (e.g. crabs, shrimps, barnacles), spiders, scorpions, mites, millipedes and centipedes.

- **artificial selection**

the process of choosing those animals that will reproduce in a given population on the basis of a given trait or traits (choosing done by humans.)

- **asexual reproduction**

reproduction without the fusion of gametes.

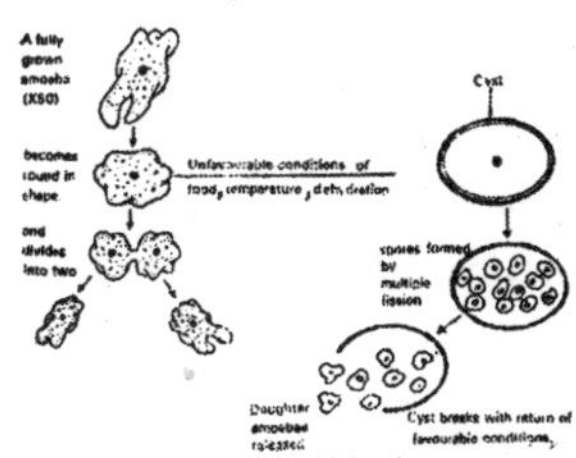

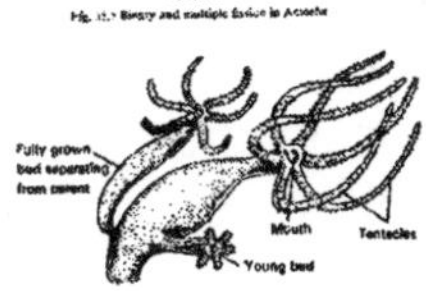

- **aspect**

the direction a slope faces. A hillside facing east has an eastern aspect.

- **ASQ (Allowable Sale Quantity)**
  the amount of timber that may be sold within a certain time period from an area of suitable land. The suitability of the land and the time period are specified in the Forest Plan.

- **assessment**
  decisions (e.g. to fight or not, to mate or not) based on the demonstrated attributes of another organism. The process by which animals evaluate the attributes of another individual, such as their fighting ability. This may involve pushing contests or bouts of high energy threat displays or outright fights.

- **assimilation efficiency**
  the percentage of energy ingested by an animal that is absorbed across the gut wall.

- **association**
  a process in learning, particularly in the type of situation originally described by Pavlov as classical conditioning.

- **associative learning**
  form of behavioural plasticity in which an animal forms an association between a stimulus that initially evoked little or no response (the conditioned stimulus) and one that normally evokes a reflex response (the unconditioned stimulus). The association is formed as a consequence of the paired presentation/occurrence of the two stimuli; related terms: Conditioning.

- **assortative mating**
  nonrandom mating on the basis of phenotype; usually used for positive assortative mating, the propensity to mate with others of like phenotypes. but may be negative assortative mating as in the rare male effect.

- **asymmetric competition**
  competition between two organisms (or species) in which one is much more severely affected than the other.

- **asymmetry**
  any incomplete identity between the two sides of the body, e.g. arms are of different lengths.

- **atavism**
  throwback inheritance. The occasional occurrence in

individuals of a species of characteristics of phylogenetically ancestral form.

- **atoll**
  a ring-shaped coral reef or string of coral islands, usually enclosing a shallow lagoon.

- **attachment**
  a tie between an animal and some place, object or social companion.

- **attack behaviour**
  a broad response class including those behaviour patterns of an animal, that when carried to completion, bring to bear one or more of the animal's effectors on the body surface of a second animal in such a way that injury of the second animal will occur if the behaviour continues.

- **attendant**
  a bird attending a nest, including parents and helpers.

- **attention**
  the notion that not all portions of incoming signals are detected equally and that by attending to one, the animal is less likely to be able to recall what occurred in the second.

- **attenuation**
  the drop in intensity of a sound with distance.

- **AUM (Animal Unit Month)**
  the quantity of forage required by one mature cow and her calf (or the equivalent, in sheep or horses, for instance) for one month.

- **aunt**
  a female that assists a dam providing care for young. In the wild the aunt is normally a close relative.

- **autapomorphy**
  a unique apomorphy; not helpful in determining species' taxonomic relationships.

- **autochthonous material**
  organic matter produced within a community (to contrast with allochthonous material that is produced outside it.)

- **autochthonous**
  formed or occurring in the place where found (opposite of allochthonous.) Behaviour that is activated by its own drive.

- **autocidal control**
  a type of pest control in which the pest is manipu-

lated so that it contributes significantly to its own control, or literally 'kills itself'.

- **autogamy**
  formation of a diploid nucleus by the union of male and female gametes in the same individual; a process of self-fertilisation.

- **autogenic succession**
  a temporal succession of species at a location that is driven by processes operating within the community in contrast with allogenic succession.

- **automimicry**
  1. in behaviour the limitation by one sex or life stage of another sex or life stage of the same species.
  2. in studies of colour mimicry, automimicry refers to the situation in which some individuals of a species store poisonous compounds (host) whereas others do not (mimic). When there is a palatability spectrum within a species, the more palatable individuals within a species protect the less palatable individuals.

- **autonomic nervous system**
  a division of the vertebrate nervous system serving internal organs such as the heart, blood vessels, lungs, intestines and also certain glands. The sympathetic nerve pathways have an emergency function and become active under stress. They have the effect of accelerating heart rate, dilating air passages to the lungs, increasing the blood supply to the muscles, reducing the activity of the intestines. The parasympathetic pathways serve a recuperative function restoring the blood supply to normal and countering the effects of the sympathetic activity. May be the origin for some displays.

- **autonomy**
  the reflexive self removal of a body part, such as the tail breaking of a lizard.

- **autopolyploid**
  a polyploid that originates through multiplying of a single set of chromosomes of the same species.

- **autoshaping**
  a type of conditioning that occurs simply because of repeated exposure to certain

stimulus contingencies. Dogs and cats quickly learn to associate the noise of a can opener with presentation of food so they come running and start drooling with the sound alone.

- **autosome**
a chromosome other than a sex chromosome.

- **autotroph**
1. an organism that is independent of outside sources for *organic* food materials and manufactures its own organic material from inorganic sources.
2. literally, 'self eater'. Organisms capable of producing their own food.

- **autotrophic succession**
a temporal succession of species at a location principally involving plants.

- **auxiliary**
a helper. See **discussion under helper**.

- **aversive stimulus**
a stimulus which, if it is applied following the occurrence of a response, decreases the strength of that response on later occurrences.

- **avoidance**
reactions that allow animals to escape from actual or potential dangers in the environment.

- **awn**
a stiff bristle, especially on the grains of cereals and grasses.

- **auxiliary bud**
a bud produced in the angle between a leaf and the stem that bears it, i.e. the normal position of a lateral bud.

- **axon**
1. elongated process for long-distance signalling. The part of the neuron that conveys information away from the cell body.
2. the portion of a nerve cell through which signals are conducted away from the cell. See **dendrite**.

- **azimuth**
horizontal direction with respect to some reference point. The sun's azimuth is its clockwise angle with respect to due south.

- **b**
1. the gain in fitness associated with performing an act.

2. the individual birth rate, the number of female offspring, one female will have per unit of time.

• **back mutation**
a second mutation that causes the allele to revert back to the original condition.

• **back propagation**
technique for training a model network of neural elements.

• **backcross**
a cross between a heterozygote and a corresponding homozygote (AA or aa).

• **background extinction rate**
normal rate of extinction — as a natural part of the evolutionary process — of various species as a result of changes in local environmental conditions and the actions of natural evolutionary forces. Extinctions not caused or contributed to by the actions of humans.

• **bacteroids**
in legume nodules the symbiotic rhizobium bacteria that have entered an active nitrogen-fixing state and have usually ceased to divide and commonly become banded and branched.

• **badge of status**
a term for features of appearance that indicate ability to maintain control of a resource, such as territory. For example, the dominance status of male sparrows (Zonotrichia querula) correlates with the area of black feathers below the beak. When the size of this bib was experimentally enlarged on some subordinate males, higher-ranking males attacked them more, showing that the wearers of badges may be tested to see whether the advertised status is merited.

• **balanced polymorphism**
two or more alleles are maintained in a population by frequency-dependent selection or heterozygous advantage.

• **balanced preference**
a preference by a consumer for food items based on the need of the consumers to obtain a balanced diet of complementary food items.

• **balancing selection**
any of several forms of natural selection leading to a balanced polymorphism.

• **Baldwin effect**
the now outmoded idea that in some instances behavioural changes could precede some more fixed evolutionary change and create a climate favouring the latter.

• **ballistic behaviour**
see **open-loop behaviour**.

• **ballooning**
the dispersal through the air of silk-producing spiders and insects suspended on silk threads.

• **bark beetle**
an insect that bores through the bark of forest trees to eat the inner bark and lay its eggs. Bark beetles are important killers of forest trees.

• **bark**
a short, sharp vocalisation produced with an open mouth as a single sound or short series of repeated sounds.

• **baroreceptor**
a sensory receptor activated by change's in barometric pressure.

• **basal area**
the area of the cross section of a tree trunk near its base, usually 4 and 1/2 feet above the ground. Basal area is a way to measure how much of a site is occupied by trees. The term basal area is often used to describe the collective basal area of trees per acre.

• **basic rank**
the paired terms 'basic rank' and 'dependent rank' are used mainly in the primate literature to signify that the positions held by two individuals in a rank order may differ according to the presence or absence of other social influences. Basic rank is the rank relationship between two animals in situations where other group members do not interfere; dependent rank is their rank status as influenced by other group members or the group as a whole.

- **basic reproductive rate**
  the average number of offspring produced by individuals in a population over the course of their life.

- **basking**
  a behaviour of reptiles and other species in which they sit out in the sun, often extending legs or feet so as to increase solar radiation.

- **batesian mimicry**
  mimicry in which a harmless or tasty species superficially resembles a dangerous or distasteful model species.

- **bathing**
  maintenance behaviour associated with dispersing water or dust through the fur or feathers so as to clean and/or remove parasites from the body surface.

- **beau geste hypothesis**
  an explanation for the evolution of multiple song types. The notion is that by singing many songs intruders will think there is more than one individual present. See **vocal mimicry**.

- **bee dances**
  a collective term referring to the various orientation behaviour patterns of the honeybee which communicate distance and direction of food to hive mates.

- **behaviour chain**
  a stereotyped sequence of stimuli and responses that can be observed repeatedly in an animal, with only minor variations in the ordinal position of each stimulus and response.

- **behaviour modification**
  behaviour modification refers to any of several techniques used to modify or eliminate a particular behaviour, e.g. desensitis-ation, extinction.

- **behaviour**
  1. the whole complex of observable, recordable, or measurable activities of a living animal. These include movements of the skeletal, cardiac, or smooth muscles, production of sound, discharge of electric organs, movements of cilia, contraction or expansion of chromatophores, glandular secretion, and changes in body chemistry (including those producing luminescence) as they are con-

cerned in the animal's interaction with its environment. 2. any parts of behaviour that are recurrently identifiable and classifiable by the observer. ('Behaviours' refer to behaviour patterns.)

- **behavioural ecology**
  a field of study which emphasizes the adaptiveness of behaviour. The study of the adaptive value of behavioural attributes of individuals in solving environmental obstacles to reproduction.

- **behavioural mimicry or ethomimicry**
  imitation of behavioural rather than morphological characteristics of other animals.

- **behavioural pattern**
  a set of responses statistically organised in time, i.e., associated together and manifesting some degree of stereotypy in the temporal sequence in which they occur.

- **behavioural physiology**
  that part of ethology that deals with the physiology of behaviour, and that part of physiology that deals with behaviour. It includes neuroethology and behavi -oural endocrinology.

- **behavioural repertoire**
  the set of behaviour patterns characteristic either of an entire species or of a single member of a species. The totality of behaviour patterns characteristic of a species.

- **behaviourism**
  1. movement in psychology that advocates the use of strict experimental procedures to study observable behaviour (or responses) in relation to the environment (or stimuli). Behaviourism deals with learning or memory under strictly controlled conditions; Related Terms: Ethology.
  2. a school of psychology: Its basic and revolutionary premise was that psychology should be regarded as the science of behaviour as opposed to the science of mental life, as William James believed. This implied that all mentalistic concepts were to be eschewed, unless they could be redefined in terms of observable behaviour and

the ascertainable or contrived circumstances of its occurrence. All psychological concepts must be reducible to those terms.

- **behaviourist**
a scientist who investigates the behaviour of animals objectively and who attempts to relate his observations together in a theoretical system that does not include concepts borrowed from introspection and mental philosophy. More specifically, a psychologist who studies behaviourism.

- **behaviours**
a vague term that usually means behavioural patterns.

- **benefit-cost analysis**
a comparison of economic benefits and costs to society of a policy, program or action.

- **benthic communities**
the plants, microorganisms and animals that inhabit the bed of aquatic environments.

- **bequest value**
the value that people place on knowing that future generations will have the option to enjoy something.

- **bergmann's rule**
within a polytypic warm-blooded species, the body size of a subspecies usually increases with decreasing mean temperatures of its habitat.

- **best-of-a-bad-job strategy**
conditional strategy of individuals where they behave according to a programmed rule making the best of the situation in which they find themselves.

- **bias**
a systematic error in sampling inherent in the sampling method.

- **bicentric distribution**
the presence of a species (or other taxonomic class) in two widely separated geographic areas.

- **big game**
large mammals, such as deer, elk and antelope that are hunted for sport.

- **bill wiping**
a bird rubs its bill on the substrate in a characteristic manner. In song birds the sides of the bill are rubbed alternatively on alternate sides of a branch whereas in

species such as chickens the sides of the bill are wiped alternately on the substrate.

- **billing**
in birds which maintain permanent pair bonds, mutual beak contract of the pair partners is common. The bills are either crossed or the bill of the arriving partner is grasped by the other as in food begging of the young.

- **bimodal population**
a population containing one or more characteristics, the measurements of which show two peaks of relative frequency.

- **binomial nomenclature**
the two-name system used to assign scientific names to all living things. Homosapiens, for example, is the scientific name for humans. The first name is the genus name and is always capitalised. This is sort of like your last name. It belongs to several of your close relatives, too, and it shows that you are all closely related. The second name is the species name is always lower case. This is like your first name, which no one else in your circle of relatives posseses and so it uniquely identifies you.

- **bioaccumulation**
an increase in the concentration of a chemical in specific organs or tissues at a level higher than would normally be expected.

- **bioacoustics**
the study of vocalisations and sound production in animals.

- **biodegradable**
1. able to be broken down into simpler substances (elements and compounds) by naturally occuring decomposers. Essentially, anything that can be ingested by an organism without causing any harm to that organism. 2. non-toxic and able to be decomposed in relatively short period even on a human time scale.

- **biodiversity**
1. in its most general sense, biodiversity refers to all aspects of variety in the living world. Spe_ifically, the term may be used to describe the number of species, the amount of genetic variation

or the number of community types present in an area.
2. the variety of life on Earth and the interconnections among living things.

A view of the Antarctic Penguines

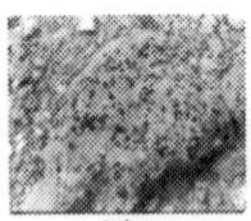
Lichens

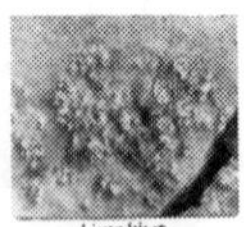
Liver Wort

Growing on rock surface around Maitri Station at Antarctica

- **biodiversity category**
the variety of biotic factors found within a specified geographic region. The combined differences of living things, generally classified in four broad categories:
1. Genetic Diversity: Variety among individuals within a species — or, more specifically, the variety in the DNA of a species.
2. Species Diversity: Variety of different organisms at the species taxonomic level.
3. Cultural Diversity: Variety of learned behaviours among individuals of a species.
4.Ecosystem Diversity: Variety of biomes and habitats occuring in the biosphere.

- **biodiversity/biological diversity**
the variety of life forms - the different plants, animals, fungi and micro-organisms, the genes they contain, and the ecosystems they form. A simple measure of the biodiversity of an area is the number of plant and animal species present. Areas of high biodiversity have many species while areas of low diversity have few species. More complex measures take into account the size of the populations of each species. Biodiversity can be considered at the levels of genetic diversity, species diversity and ecosystem diversity. Genetic diversity refers to the variation of genes within species. Species diversity refers to the variety of species. Ecosystem diversity encompasses the broad differences between ecosystem types. It also looks at the diversity of habitats and ecological processes occurring within each ecosystem type. Biodiversity is not merely an aesthetic component of the environment. It pro-

vides ecosystems with greater resilience to recover and cope with stress caused by things such as drought or human-induced habitat changes. It is argued to be the main source of the world's present and future renewable resources. The protection and maintenance of biodiversity is therefore regarded as one of the fundamental responsibilities.

- **biogenetic law**
  haeckel's term for the principle that ontogeny recapitulates phylogeny.

- **biogeochemical cycling**
  the movement of chemical elements between organisms and non-living compartments of atmosphere, lithosphere and hydrosphere.

- **biogeography**
  1. the study of the geographical distribution of organisms.
  2. the study of living systems and their distribution. Biogeography is important to the study of the Earth's biodiversity because it helps with understanding where animals and plants live, where they don't, and why.
  3. branch of biology that deals with the geographic distribution of plants and animals.

- **biological clock**
  a physiological mechanism for measuring time or maintaining endogenous rhythms.

- **biological control**
  1. the use of a pest's natural enemies in order to control that pest.
  2. the use of natural means to control unwanted pests. Examples include introduced or naturally occurring predators such as wasps, or hormones that inhibit the reproduction of pests. Biological controls can sometimes be alternatives to mechanical or chemical means.

- **biological diversity**
  the number and abundance of species found within a common environment. This includes the variety of genes, species, ecosystems, and the ecological processes that connect everything in a common environment.

- **biological oxygen demand**
  the rate at which oxygen disappears from a sample of water - a measure of deoxygenating ability commonly used as an index of the quality of sewage effluent.

- **biological pesticides**
  a preparation used to provide immediate control of a pest and which consists of biological as opposed to chemical material.

- **biological productivity**
  the intensity of life form production in an ecosystem or part of an ecosystem.

- **biological species concept**
  a concept of the species category stressing reproductive isolation and the possession of a genetic program effecting such isolation. A concept of the category species based on the reproductive isolation of the constituent population from other species.

- **biomagnification**
  the increasing concentration of a compound in the tissues of organisms as the compound passes along a food chain, resulting from the accumulation of the compound at each trophic level prior to its consumption by organisms at the next trophic level.

- **biomass**
  1. the weight of living material. Most commonly used as a measure per unit area of land or per volume of water. Commonly includes the dead parts of living organisms, e.g. the bark and heart wood of trees and the hair, claws etc. of animals which are strictly 'necromass'.
  2. the total weight of all living organisms in a biological community.

- **biome**
  1. one of the major categories of the world's distinctive plant assemblages, e.g. the tundra biome, the tropical rainforest biome.
  2. the complex of living communities maintained by the climate of a region and characterised by a distinctive type of vegetation. Example of biomes in North America include the tundra, desert, prairie and the western coniferous forests.

3. a specific type of terrestrial region inhabited by well-defined types of life, especially zones of vegetation, that generally cannot live outside that specific region. Examples include types of deserts ('high desert' like the Mojave or 'low desert' like the Chihuahua), grasslands (prairies, coastal dunes), and forests (lodgepole pine vs. taiga; temperate rain forest; bamboo forest, tropical rain forest, cloud forest, etc.).

- **bionomics**
  See **ecology**.

- **biorational insecticides**
  insecticides which have no, or relatively limited, adverse effects on other, non-pest organisms in the pest's environment.

- **biosphere**
  1. the portion of the earth and its atmosphere in which living organisms exist or that is capable of supporting life.
  2. all of earth's ecosystems combined into one inclusive unit. Also called the ecosphere.
  3. the living organisms and their environment composing the biosphere. All life on earth and the realms that support it, from the outermost reaches of the atmosphere to the deepest trenches of the seas.

- **biota**
  1. the fauna and flora together; all the living organisms at a location.
  2. the plant and animal life of a particular region.

- **biotic**
  1. living; usually applied to the biological aspects of an organism's environment, i.e. the influences of other organisms.
  2. living. Green plants and soil microorganisms are biotic components of ecosystems.
  3. refers to the living components of the environment (such as plants, animals, and fungi) that affect ecological functions.

- **biotrophic**
  a word used of parasites that can complete their development on only a living host.

- **biotype**
  a physiologic race or a group of individuals having distinctive genetic characters in common.

- **bipolar neuron**
  neuron with two dendrites, one extending from each end of the cell body. This structure is typical of many invertebrate sensory neurons.

- **bipolar**
  neurons that conduct in both directions.

- **bivoltine**
  a life-history pattern with two generations in one year. Partially bivoltine refers to a life-history pattern in which animals sometimes produce offspring that over winter (i.e. one generation per year) and sometimes produce offspring that emerge later in the same season (i.e. bivoltine).

- **bivouac**
  the mass of army ants within which the queen and brood find refuge.

- **black box approach**
  an approach to the study of mechanisms that results in developing hypotheses about internal patterns and structure based on altering sensory (or other) input and observing the consequent changes in behaviour.

- **blending inheritance**
  before the time of Mendel, the belief that inheritance was not particulate.

- **block diagram**
  a diagram of a system in which arrows represent variables and blocks represent parameters of the system.

- **blue-green algae**
  cyanobacteria.

- **BMP (Best Management Practices)**
  practices designed to prevent or reduce water pollution.

- **board foot**
  a measurement term for lumber or timber. It is the amount of wood contained in an unfinished board 1 inch thick, 12 inches long and 12 inches wide.

- **bond**
  social attachment; a specific dependence relationship between two or more individuals.

- **bootstrap method**
a technique whereby an unknown distribution is estimated by repeated sampling from a sample distribution (sampling using the computer) to arrive at confidence intervals for that distribution with which your observed results can be then compared.

- **boreal**
1. northern.
2. pertaining to the north.

- **boredom**
prolonged exposure to repetitive sights, sounds, etc., leads to impaired performance in tasks which involve attention to stimuli. It is not possible to know whether animals have the same subjective experience as humans because it cannot be measured and thus this term should be avoided when referring to animals.

- **boring**
refers to organisms capable of penetrating solid substrate by mechanical scraping or chemical dissolution.

- **botanical insecticides**
chemicals extracted from plants for the control of insect pests.

- **bottleneck effect**
occurrence of genetic reorganisation of a population through genetic drift which is caused by a drastic decrease in abundance.

- **boundary layer**
the relatively still layer of water just above the bed of a river.

- **bout**
a bout of behaviour occurs when an action is repeated several times in quick succession, with longer intervals between, i.e. inter-bout intervals.

- **bower**
display structure of bower birds. Construction by males for mate attraction.

- **brachiation, brachiating**
progression through trees by swinging from branch to branch with the forelimbs.

- **brackish**
1. saline water with a concentration between freshwater and seawater.

2. slightly salty or briny. Brackish water is saltier than fresh water but less salty than seawater.

- **bradytely**
evolving at a relatively slow rate.

- **breathing**
the movement of air in and out of lungs and thus a part of the total function of gas exchange.

- **breeding population size**
the number of individuals in a population actually involved in the reproduction of a given generation.

- **bridling**
the post-copulatory display of a male anatid duck.

- **broadcast burn**
a prescribed fire that burns a designated area. These controlled fires can reduce wildfire hazards, improve forage for wildlife and livestock, or encourage successful regeneration of trees.

- **broadcast promiscuity**
a breeding system in which the gametes are released into the environment and fertilised by males in the surrounding water.

- **broken-wing display**
the predator is distracted by the adult bird which looks as though it is feigning some locomotory impairment such as limping or wing injury.

- **brontophobia**
fear of thunder.

- **brood parasitism**
the act of leaving eggs or progeny to be reared by an individual that is not the parent - usually a member of another species.

- **brood parasitism**
the laying of eggs in the nest of another individual either a conspecific or a heterospecific and then allowing the host to rear them. Also called 'cuckoo behaviour'.

- **brood patches**
in birds there are areas of ventral skin that become defeathered, vascularised and edematous during incubation.

- **broodiness**
a part of maternal behaviour directed specifically to

the hatchlings, most commonly associated with birds. The behaviour includes protecting and covering the chicks.

- **brooding**
in birds, the parents sitting or crouching over young to provide warmth and cover.

- **browse**
twigs, leaves and young shoots of trees and shrubs that animals eat. Browse is often used to refer to the shrubs eaten by big game, such as elk and deer.

- **browsers**
vertebrate herbivores that feed from trees or shrubs.

- **bruce effect**
a condition in which the presence of an unfamiliar male prevents the implantation of a fertilised egg or induces abortion.

- **buffer**
a land area that is designated to block or absorb unwanted impacts to the area beyond the buffer. Buffer strips along a trail could block views that may be undesirable. Buffers may be set aside next to wildlife habitat to reduce abrupt change to the habitat.

- **buller steer syndrome**
bullers are steers that stand for mounting by other steers called riders. The behaviour can cause injuries or exhaustion to the animals being ridden. May be related to crowded conditions stress and hormone implantation.

- **bulling**
mounting and chin-pressing of a cow by another cow, usually when the instigator is close to or in estrus.

- **burying food**
domestic carnivores commonly bury food, a behaviour commonly found in the wild to keep the food away from scavengers and conspecifics.

- **butting**
a distance-increasing behaviour shown by ungulates as they aggressively strike with the head.

- **bycatch**
fishes, or any other animals, that are accidentally caught in fishing gear.

- **by-product mutualism**
  a mutualism in which benefit to neighbours is an incidental by-product of actions taken by the donor for its own benefit. Reciprocation is not required to maintain by-product mutualism, as it is for score-keeping mutualisms.

- **c**
  opportunity cost. The gain in fitness from performing the alternative act.

- **cable logging**
  logging that involves the transport of logs from stump to collection points by means of suspended steel cables. Cable logging reduces the need for the construction of logging roads.

- **cable properties**
  passive conducting properties of axons and dendrites i.e. when the cell membrane is passive an axon or dendrite behaves electrically like an insulated cable.

- **caching**
  the laying by of food for use during times of insufficient supply. Also called food storage.

- **caecum**
  a blindly ending sac at the junction between the small and large intestines.

- **cainism**
  the killing of siblings, fratricide, siblicide.

- **calcareous**
  composed of, or containing lime or limestone.

- **call**
  vocalication that is simpler than passerine song.

- **cambrian**
  the earliest geologic period of the Paleozoic Era.

- **camouflage**
  a form of visual deception by means of which an animal can elude predators or a predator may lurk undetected awaiting a suitable prey.

- **canalisation**
  a property of developmental pathways whereby a standard phenotype is achieved in spite of genetic or environmental disturbances. The operation of internal factors during development that reduce the effect of perturbing influences, thereby constraining variation in the

phenotype around one or more modes.

- **canalising selection**
  the selection of genes that would stabilise the developmental pathways so as to make the phenotype less susceptible to the effects of genetic or environmental disturbances.

- **cannibalism**
  the consumption of conspecifics as food.

- **canopy**
  the part of any stand of trees represented by the tree crowns. It usually refers to the uppermost layer of foliage, but it can be use to describe lower layers in a multi-storied forest.

- **capacitance**
  property of the cell membrane that allows an electrical charge to be stored. It measures the amount of charge that needs to be transferred between conductors for setting up a given potential difference. Capacitance introduces distortion in the time course of passively conducted signals as this storage takes time to get charged. Capacitance thus slows down a voltage response by a characteristic time factor.

- **capture - recapture**
  a method for estimating the size of populations of mobile organisms (usually animals), in which one or more samples are captured, marked and released, and one or more samples examined for the proportion of recaptured marks (broadly speaking, high in small populations and low in large.)

- **carbohydrase**
  an enzyme that acts on carbohydrate.

- **carbon c-3 plants**
  those in which the assimilation of atmospheric carbon dioxide is directly via the enzyme ribulose-1,5-bisphosphate carboxylase

in the cells of the leaf mesophyll.

- **carbon c-4 plants**
species of higher plants in which the assimilation of atmospheric carbon dioxide in the photosynthetic process is indirect, via the enzyme phosphoenol pyruvate carboxylase in the sheaths surrounding the veins of the leaves.

- **carboniferous period**
geological period from *c.* 270 to 220 million years ago.

- **care-giving behaviour**
behaviour of guarding and protecting young, not necessarily your own.

- **carnivory**
the consumption by an organism of living animals or parts of living animals.

- **Carolus Linnaeus**
Swedish botanist and originator of the binomial nomenclature system of taxonomic classification.

- **carrying capacity**
1. the maximum population size that can be supported indefinitely by a given environment, at which intraspecific competition has reduced the per capita net rate of increase to zero. An idealized concept not to be taken literally in practice.
2. the limit to an environm-ent's capacity to withstand use or activity. Beyond this limit the activity's impact (social, ecological or both) is unacceptable.
3. this term was originally used in agriculture and ecology. In agriculture it is the maximum number of grazing animals that an area can continue to support without deterioration of the soil and vegetation cover. Carrying capacity can also refer to social or economic capacity.
4. environment's maximum persistently supportable load.
5. the amount of animal or plant life (or industry) that can be supported indefinitely on available resources; the number of individuals that the resources of a habitat can support. Also called biological carrying capacity.
6. the largest population size that can be maintained by an environment.

- **caste polyethism**
  the division of labour roles within a social insect colony.
- **caste**
  found in social insects, where some individuals do all the reproducing (queens and kings) whereas others act as non-reproductive workers. In ants caste refers to morphological differences among workers.
- **catabolism**
  the decomposition by living organisms of complex organic molecules to simpler forms, with the release of energy.
- **catastrophe**
  a major change in the environment that causes extensive damage and usually widespread death and occurs so infrequently that the effects of natural selection by similar events in the past (if they have ever occurred) do not remain in the 'genetic memory' of the species. The Mt. St. Helens volcanic eruption was in this sense a catastrophe.
- **catastrophism**
  Cuvier's explanation of the successive replacements of fossil faunas as the result of geological catastrophes.
- **catch per unit effort**
  usually applied to the harvesting of a natural resource (e.g. marine fish), the total catch (in terms of numbers or biomass) divided by the total harvesting effort (e.g. a product of the total number and size of ships and the number of days that they fished).
- **category**
  in taxonomy, one of the ranks (e.g., genus, family) of classification. See **taxon**.
- **caterwauling**
  vocalistion by cats, particularly intact tomcats to advertise their territory to potential mates and to warn intruders away.
- **causal factor space**
  the array of motivational variables that influence the likelihood that an action will be performed.
- **cavity**
  a hole in a tree often used by wildlife species, usually

birds, for nesting, roosting and reproduction.

- **cell body**
  region of the neuron that includes the nucleus and associated intracellular organelles.
- **cellulolytic enzymes**
  enzymes that act on cellulose.
- **cellulose**
  a complex polymer of glucose molecules. The fundamental cell wall constituent in all green plants.

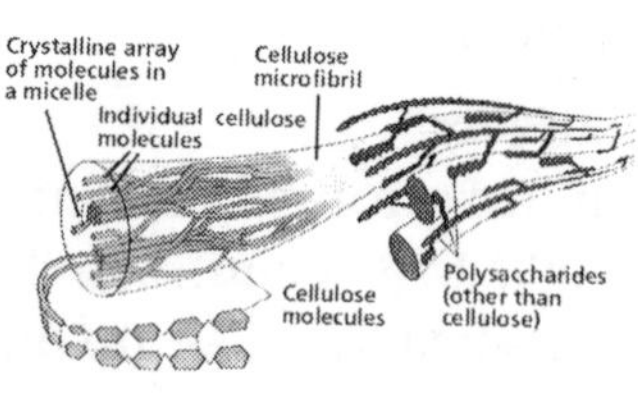

- **census**
  in ecology, an attempt to count every member of a population.
- **centre**
  1. a locus in the nervous system characterised by the presence of a number of cell bodies and synapses. The excitation of which by appropriately specified electrical stimulation, may yield discrete motor or autonomic behaviour patterns and experimental destruction of which is followed by the disappearance or gross modification of discrete motor or autonomic behaviour pattern.
  2. a functionally coordinated, but not necessarily localised, group of neural structures having the properties of a centre.
  3. a hypothetical neural structure, or place or set of places in the central nervous system, of unspecified anatomical properties, which is presumed to act as a unit upon excitation by another such place or other such places by sending nerve impulses that govern the occurrence of some innate response.
- **centre-surround**
  connectivity of a receptive field in which stimuli elicit the largest response in a central region with an opposite polarity in the surrounding region.
- **central nervous system**
  main organ of the nervous system. In vertebrates, it is made up of the brain and the spinal cord.

- **central pattern generator**
 1. group of neurons in the central nervous system that, due to the properties of the neurons and the connections between them, can produce a specific pattern (usually rhythmic) of output in the absence of sensory feedback.
 2. a group of cells in the central nervous system that can produce the particular pattern of signals needed to cause a fully functional behavioural pattern.

- **centromere**
 the portion of a chromosome, usually greatly restricted in length, to which spindle fibres are attached in cell division.

- **chain of response**
 behaviour sequence in which each movement or activity brings an animal into a situation where the next action pattern is evoked.

- **chain reflex**
 Pavlov maintained that all behaviour could be explained as a sequence of conditioned and unconditioned reflexes and that complex behavioural responses were simply put together in chains.

- **challenge display**
 a high intensity aggressive display performed by a male to a conspecific male.

- **change of function**
 an evolutionary phenomenon whereby an organ which ceases its original function may take over a new function.

- **channel**
 opening or 'pore' in a membrane through which ions or molecules can move.

- **chaos**
 applied to a time-series of population abundance or density, a pattern in which elements of the pattern are never repeated exactly, and where two very similar abundances in the time-series follow trajectories subsequently which diverge from one another exponentially. Nonetheless, the series is not random and fluctuates within definable limits.

- **chaparral**
a thicket of low evergreen oaks or dense tangled bushwood.

- **character displacement**
1. a measurable physical difference between two species which has arisen by natural selection as a result of the selection pressures on one or both from competition with the other.
2. a divergence of equivalent characters in sympatric species resulting from the selective effects of competition.

- **character polarity**
the direction of character state change; determined by out group comparison.

- **character state**
one of several variant.

- **character**
an attribute of a member of a population or taxon by which it differs from a member of a different group or taxon, more generally a feature or trait.

- **chemical control**
the use of pesticides and herbicides to control pests and undesirable plant species.

- **chemical synapse**
type of synapse in which communication is achieved by release of chemicals from the pre-synaptic membrane, diffusion across the synaptic cleft, and binding to receptors on the post-synaptic membrane.

- **chemoreceptor**
1. sensory receptor that responds to binding of a specific class of molecule.
2. sensory cells or sense organs for the perception of chemical stimuli; primarily the cells or organs of smell and taste.

- **chemosynthesis**
the synthesis of organic molecules by certain bacteria that use the energy released by specific inorganic molecules.

- **chemotaxis**
an orientational reaction by free-moving animals induced by a chemical stimulus, smell or taste.

- **chiasma**
(plural: chiasmata). An attachment between homologous chromosomes in mitosis or meiosis; thought to be the

point where physical exchanges between homologues occur, i.e. crossing-over.

- **chimera**
an individual whose cells are not all of the same genotype.

- **chin raising**
epigamic behaviour in ducks in which both partners face one another and together raise the bill angle above the horizontal.

- **chlorophyll**
green pigment(s) found in almost all plants and playing a crucial role in the capture of radiant energy in the process of photosynthesis.

- **chloroplast**
the inclusions (plastids) within plant cells which contain chlorophyll.

- **choice (as in female choice)**
there are two types of choice,- active and passive. In active choice the female compares the alternatives and acts upon that information. Passive choice, on the other hand, occurs simply when females mate non-randomly.

- **choice points**
1. that position in a T-maze, or other maze, or on a discrimination apparatus, from which it is possible for the animal to give only one of two or more alternative responses.
2. in a sequence of behaviour, the point where alternative actions can occur.

- **choice test**
1. a type of tests for determining the learning capacity or sensory abilities of animals. Certain colours, forms, patterns, etc. of objects are presented and the animal must make a choice between the correct one, for which it is rewarded, and the wrong one for which it is given no reward or punished.
2. a test of the decision-making rules of individuals.

3. a method for determining the social preferences of animals and for investigating their learning and discrimination capacities and the releasing and directing effects of stimuli. A choice test presents the experimental animal with two or more objects or live animals, for example on a selection platform or at opposite ends of a multipartitioned cage or aquarium. The animal's reactions or the time it spends in the vicinity of each option are recorded. As a rule the choice objects are presented together (simultaneous choice test), but sometimes they are presented one after the other (successive choice test). The results of simultaneous and successive tests do not always agree. When presented with comparable calls of neighbouring adults and parents, chicks recognise their parents' calls in simultaneous presentations, but showed no discrimination in successive presentations. In the natural situation chicks of this age are better off with their parents, but if they get separated from their parents, their only hope of survival is to be adopted into another family, so indiscriminate responsiveness is then the more adaptive course.

- **choking**
  a display pattern of gulls in which the members of a pair lower their breasts, bend the legs, nod their heads and rhythmically jerk the head as if to peck at the ground, although they do not reach the ground.

- **chorusing**
  communal song; song of aggregated animals.

- **chromatid**
  a visible subdivision of a chromosome having but one centromere (or behaving as if the centromere were undivided); a half chromosome.

- **chromosome**
  a threadlike structure present in the nucleus of cells and containing genes in linear arrangements.

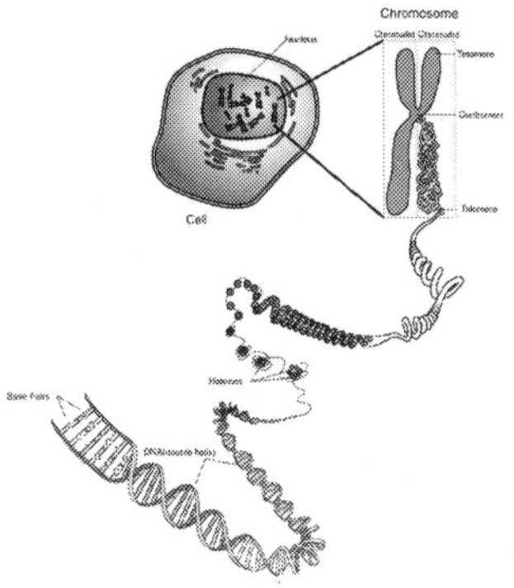

- **chronospecies**

segments of a single lineage, between which there has been gradual evolutionary change so that the descendants are morphologically distinctive but for which any subdivision within the grad-uational series would be arbitrary. A segment of an evolving lineage preserved in the fossil record, that differs enough from earlier or later members of the lineage to be given a different binomial (name). Not equivalent to biological species.

- **circadian clock**

inherent timekeeping mechanism with a capacity to drive or coordinate a circadian rhythm.

- **circadian rhythm**

1. biological rhythms with periodicity of approximately 24 hours.
2. the daily pattern of activity of a cycle that is approximately one day in length expressed independent of environmental changes.

- **circannual rhythm**

changes in behaviour that recur at 12-month intervals that expresses itself independent of environmental changes.

- **circular overlap**

the phenomenon in which a chain of contiguous and intergrading populations of a species, where the terminal links overlap with each other and the populations behave like good (non-interbreeding) species.

- **cistron**

a DNA fragment or portion that specifies or codes for a particular polypeptide.

- **clade**

a species or set of species representing a distinct branch of a phylogenetic tree, i.e. descended from a common ancestor.

- **cladism**
school of taxonomy that bases its classification or evidence on common ancestry through shared, derived characters.

- **cladistic**
pertaining to branching patterns; a cladistic classification classifies organisms on the basis of the historical sequences by which they have diverged from common ancestors.

- **cladogenesis**
the splitting or branching of one lineage into two during evolution.

- **cladogram**
1. a phylogenetic tree emphasizing branching patterns and when they occur. See **phenogram**.
2. A diagram showing the branching points of divergence over the course of evolution.

- **clandestine evolution**
evolutionary change introduced and developed in juvenile stages and incorporated into descendant adult stages by paedomorphosis. Since juvenile stages fossilize so rarely, these evolutionary events go unnoticed until they are promoted to a descendant adult.

- **claspers**
structures found in frogs and toads, sharks, horseshoe crabs, some insects and many other species for holding on to the female.

- **classical conditioning**
1. behaviour changes when an organism comes to associate one stimulus with another, a reflexive or automatic response transfers from one stimulus to another. For instance, a person who has had painful experiences at the dentist's office may become fearful at just the sight of the dentist's office building. In classical conditioning the conditioned and unconditioned stimuli are presented regardless of what the learner does.
2. the experimental procedure of repeatedly presenting the animal with a stimulus (unconditioned stimulus or US) of some reflex contiguously or almost contiguously in time with a neutral stimulus (conditioned stimulus or CS). Con-

ditioning is said to ha[illegible] c-curred if, and only if, [illegible] e-sponse (condition[illegible] re-sponse or CR) similar to, but not necessarily identical with, that of the reflex can later be elicited by the initially indifferent stimulus.

- **classification**
a mathematical procedure for categorising communities in which communities with similar species compositions are grouped together in subsets.

- **claw-sharpening**
cats claw objects as a part of grooming, marking and stretching. The behaviour of extending fore claws into soft wood or fabric helps remove fragments of claw that are being shed and keeps the claws sharp. There is also an olfactory message left by sweat glands on the feet and a visual cue left by changes in the surface with repeated claw-sharpening bouts at the same location.

- **claw-waving display**
in fiddler crabs the large cheliped is used in threatening conspecifics as well as attracting females. Also found in other crab species.

- **cleaner fish mimic**
the sabre-toothed blenny has copied the form, coloration and locomotory behaviour of the true cleaner fish and profits from the mimicry by biting small pieces out of the skin, fins and gills of the fish expecting to be cleaned.

- **cleaning symbiosis**
an interspecific, mutualistic interaction in which animals are specialised for removing parasites and necrotic tissue from the body surface of another species. The evolution of cleaning is associated with a set of signals which permits interspecific communication. Most common in fish and shrimp. A form of association between animals of different species that benefits them both. The requirements of one partner is to be rid of ectoparasites; for the other the ectoparasites constitute an important or even the sole source of food.

- **clear cut**

a harvest in which all or almost all of the trees are removed in one cutting.

- **cleistogamy**

self-pollination within a flower that does not open.

- **cleptoparasitism**

in entomology this refers to brood parasitism. A parasitic fly for example, lays its egg on the provisions of a wasp. The fly larvae develop more quickly and either eat the host lava or out compete it.

- **clever hans**

the horse named Clever Hans was reputed to be able to do arithmetic; when presented with a sum written on a blackboard, it gave the answer by tapping out the number with its foot. Tests proved that the animal was really responding to subtle clues, unwittingly given by the questioner, that signalled when to stop tapping. Ever since this demonstration proved how unconscious cueing can affect animal intelligence; behavioural scientists have been alert to the possibility and tried to exclude it. Some of the recent work on language learning in apes has been criticised on the grounds that a Clever Hans effect might have been involved.

- **climax**

1. the presumed endpoint of a successional sequence; a community that has reached a steady state.
2. the culminating stage in plant succession for a given site. Climax vegetation is stable, self-maintaining and self-reproducing.

- **climax community**

a community capable of perpetuation under the prevailing climatic and edaphic conditions.

- **clinal speciation**

a kind of parapatric speciation in which reproductive isolation develops along a cline.

- **clinch position**

body position of two fighting adversaries character -ised by mutual insertion of the head between the hind legs of the opponent.

- **cline**

a gradual change in an allele frequency or in the mean of a character over a geographic transect.

• **clock-shift experiment**
a way of studying compass reactions of animals when those reactions are time compensated. The idea is that by altering the time of day during which the animal is active (i.e. giving the animal jet lag), you can make predictions about which stimuli the animal is responding to. For example, if the animal is using a sun compass and it is always fed in the south end of its cage, then if its time clock is shifted back by three hours, the animal will go to what it thinks is the south end of the cage, i.e. at noon it will go toward the sun, but since it is clock-shifted by three hours it will actually go to the west of south by 45 degrees (15 degrees per hour).

• **clonal dispersal**
the movement or growth away from each other of the parts of a modular organism (commonly, but not necessarily, becoming detached from each other).

• **clone**
1. the whole product of growth from a single zygote, in organisms that grow by the repeated iteration of units of structure (modular organisms) and in which the units have, at least potentially, the capacity for physiological independence. The parts of a clone are genetically identical except for what are probably rare somatic mutations.
2. a lineage of individuals produced asexually by parthenogenesis or vegetative reproduction by mitotic division.

• **closed population**
a population with no genetic input other than by mutation, i.e. with no gene flow.

• **closed-loop behaviour**
1. behaviour that is performed with feedback control. See **feedback**.
2. the distribution of organisms in which individuals are closer together than if they were distributed at random or equidistant from each other.

• **clutch**
a set of eggs laid on consecutive days.

- **coacervates**
aggregates of varying degrees of complexity resulting from the interaction of two or more colloids.

- **coadaptation**
the harmonic, epistatic interactions of genes brought together by natural selection.

- **coarse filter management**
land management that addresses the needs of all associated species, communities, environments, and ecological processes in a land area. See **fine filter management**.

- **coarse grained environment**
the environment experienced by an organism with poor homeostatic mechanisms, low behavioural plasticity and mobility; experienced as sets of alternative, disconnected conditions.

- **coastcare**
1. it provides the opportunity for communities to become involved with coastal management involving protection, rehabilitation or enhancement of the natural and cultural environment. 2. coastcare encourages the involvement of and sponsorship by local coastal industries and businesses. It is a joint Commonwealth, state and local government program. Coastcare is delivered through Coast Action in Victoria and Coastwest in Western Australia.

- **coastnet**
a forum on the Internet for communicating about coastal management issues. Discussion groups discuss, debate, exchange information and share expertise on a variety of topics.

- **coding**
means by which information about sensory quality or strength is conveyed to the central nervous system, See also **Frequency Code**.

- **codon**
a group of three nucleotides that code for an amino acid.

- **coefficient of interference**
a measure of the extent to which interference amongst consumers increasingly depresses individual consumption rates as consumer density increases.

- **coefficient of relatedness**

a measure of genes shared as a result of having a common ancestor; the probability that an allele present in one individual will be present in a relative; the proportion of the total genotype of one individual present in the other as a result of shared ancestry.

- **coefficient of variation**

1. a statistical term, referring to the standard deviation of a distribution divided by the distribution's mean, and hence providing a standardized measure of the variation in a distribution, which does not increase simply because the mean itself increases or because the units of measurement change.

2. the standard deviation divided by the mean, multiplied by 100. Used for comparing variation in a character when the means are different.

- **coelom**

a 'true' body cavity in animals, filled with fluid, lined with mesoderm and communicating with the exterior by special ducts.

- **coevolution**

1. the process by which members of two (or more) species contribute reciprocally to the forces of natural selection that they exert on each other, e.g. parasites and their hosts.

2. the interdependent evolution of two or more species having an obvious ecological relationship.

3. the evolution of two or more species closely associated ecologically, with each exerting significant selective pressures on the other.

- **coexistence**

the living together of two species (or organisms) in the same habitat, such that neither tends to be eliminated by the other. This begs lots of questions about the meaning of 'living together' and the 'same habitat'.

- **cognition**

1. a hypothetical stimulus-association or perceptual organisation postulated to account for expectancies.

2. as used in animal behaviour studies of cognition generally refer to mental processes that are pre-

sumed to be occurring within the animal, but which cannot be directly observed.
3. cognition applies to an individual's mental functions including the ability to learn, remember and think.

- **cognitive ethology**
that part of ethology having to do with animal cognition.

- **cognitive map**
an internal representation of the layout of an area of the environment. Such representation is implied by the ways rats can run mazes when presented with shortcut or detours they have not previously taken.

- **cohort generation time**
an approximation to the true generation length, which takes no account of the fact that some offspring may themselves develop and give birth during the reproductive life of the parent; hence, simply the average time between the birth of a parent and the birth of its offspring.

- **cohort life table**
a life table constructed by monitoring a group of individuals all born during the same short period, from this time of birth through to the death of the last surviving individual.

- **cohort**
those members of a population that are of the same age.

- **coital lock**
powerful contraction of the vulva around the bulb of an inserted penis which temporarily prevents separation of copulating partners, e.g. in copulating dogs.

- **collateral kin**
kin who lie on different branches of a genealogy below a bifurcation, as opposed to lineal kin, who lie on the same line of descent or branch. Parents and offspring are lineal kin. Siblings are collateral kin of each other.

- **collector - filterers**
aquatic animals that filter small particles of organic matter from the water flowing over them.

- **collector - gatherers**
aquatic animals that gather small particles of organic matter from the sediment.

- **collector roads**
these roads serve small land areas and are usually connected to a forest system road, a county road, or a state highway.

- **colloid**
a dispersion of one substance within another, having properties of both a solution and a suspension.

- **colonisation**
the entry and spread of a species (or genes) into an area, habitat or population from which it was absent.

- **colony**
1. in biology, a congregation of individuals of the same species, 'congregation' here carrying no implication about the nature of the cohesion.
2. individual members may be joined together as branching growth, in which case their individuality may appear to be merged in the whole, as in many of the marine animal forms described as colonies (for example, hydroid colonies, Portuguese men-of-war, sea pens). They may be held together by gelatinous secretions, as in the colonies of many algae.
3. colonies can be free-living, as the breeding colonies of many birds are, and the societies (sometimes described as colonies) of ants and termites, and sedentary groups of spiders. In the case of insects, however, because of the degree of mutual dependence between individuals, the question can be raised of the extent to which they have true individuality.

- **colour vision**
refers to the ability of animals to detect the wavelengths of light as opposed to their relative shades of grey.

- **command neuron**
a cell in the CNA that controls the activation of a behaviour.

- **command neurons**
cells, when activated, are capable of producing complex behaviour patterns in the absence of any meaningful external stimuli. These are both necessary and sufficient to control the onset, duration, and vigour

of a complex behaviour pattern without sensory input.

- **commensalism**
an interaction in which one organism (or species) beneficially affects a second organism (or species), but the second has no effect (good or bad) on the first.
2. an ecological relationship between species in which one is benefited but the other is little affected.

- **communal breeding**
1. a system of breeding that is characterised by the normal presence of helpers at some or all nests. Often this results in the presence of three or more birds attending the young at a given nest, though this need not be true when a parent is missing. Brood capture, in which a parent takes over the care of young not its own by forcing the real parents to abandon them, which happens in some waterfowl, is excluded by tradition though not by the above definition.
2. cases in which individuals care for young not their own as a result of deception through intraspecific or interspecific brood parasitism are also excluded by tradition. In these cases the actual parents (the parasites) do not care for their own young. In this way brood parasitism differs from communal (cooperative) breeding.

- **communal courtship**
displays associated with group mating.

- **communication**
1. signal transmission and information transmission and information transfer as a prerequisite of social behaviour.
2. an animal is said to have communicated with another animal when it can be shown to have influenced its behaviour.
3. animal A is said to have communicated with B when A's behaviour manipulates B's sense organs in such a way that B's behaviour is changed. The cooperative transfer of information from a signaller to a receiver.
4. intentional information transfer between sender and receiver. It is usually of advantage to both.

- **community**

1. the species that occur together in space and time.
2. a characteristic group of plants and animals living and interacting with one another in a specific region under similar environmental conditions.
3. the group of organisms found in a particular place.

- **community consultation**

also referred to as 'public consultation'. These terms are used for procedures whereby members of the community can express their views about a proposed development, and the decisions about the development should take these views into account. Unlike community participation, it is used in situations where members of the community are not playing a role in the actual decision-making.

- **community participation**

procedures whereby members of a community participate directly in decision-making about developments that may affect the community.

- **community stability**

the tendency of a community to return to its original state after a disturbance.

- **community structure**

the list of species and their relative abundances in a community.

- **comparative method**

a procedure for testing hypotheses on the adaptive value of a trait using disciplined comparisons among species.

- **comparative study**

in ethology comparative studies are those which compare behaviour among two or more species and allow one to determine either behavioural homologies or convergence.

- **compartmentalisation in communities**

a tendency in communities to be organised into subunits within which interactions are strong but between which interactions are weak.

- **compass**

1. mechanisms that allow you to find a particular spatial direction.

2. an animal is said to have a compass when it is able to travel in and maintain a fixed direction.

- **compass reactions**
transverse orientation at a temporarily fixed angle to the direction of the external stimulus.

- **compensating variation**
the amount of money that leaves a person as well off as they were before a change. Thus, it measures the amount of money required to maintain a person's satisfaction, or economic welfare, at the level it was before the change.

- **compensation point**
the intensity of radiation at which photosynthesis equals and balances respiration.

- **competition**
1. an interaction between two (or more) organisms (or species), in which, for each, the birth and/or growth rates are depressed and/or the death rate increased by the other organisms (or species).
2. use of the same limited resources by two or more organisms.

- **competition coefficient**
in interspecific competition, a measure of the competitive effect of one species on another relative to the competitive effect of the second species on itself.

- **competitive exclusion**
1. the elimination from an area or habitat of one species by another through interspecific competition.
2. the principle that no two species can coexist at the same locality if ecological requirements are identical.

- **competitive release**
the expansion of the niche of a species associated with the lack of competition with other species; for instance, because it occurs on an island where competitors are absent.

- **complementarity**
a condition of two resources of which less is required when they are taken together than when consumed separately.

- **complementary goods**
goods that are often purchased together, such as bread and butter.

- **complementary resources**
a pair of resources for which consumption by the consumer of one resource reduces the consumer's requirements for the other.

- **composition**
what an ecosystem is composed of. Composition could include water, minerals, trees, snags, wildlife, soil, microorganisms and certain plant species.

- **concerted evolution**
maintenance of a homogeneous nucleotide sequence among the members of a gene family (q.v.), which sequence evolves over time.

- **concorde fallacy**
the principle that the amount of time and energy an animal expends on a project in the past dictates its future investment.

- **conditional strategy**
1. a set of rules that enables individuals to exhibit different tactics under different environmental conditions. 2. the capacity to be behaviourally flexible. When the behaviour shown by an animal depends on circumstances, such as age or size.

- **Conditioned Response (CR)**
a response which appears or is modified as a consequence of conditioning.
1. backward conditioned response, the conditioned response that is set up by classical conditioning when the unconditioned stimulus precedes the conditioned stimulus.
2. delayed conditioned response, the conditioned response that is set up by classical conditioning.

- **conditioned stimulus**
1. in classical conditioning a stimulus which originally does not evoke any response similar to the unconditioned response, but which during conditioning acquires the property of eliciting this response or a similar one.
2. the original neutral stimulus.

- **conditioning**
1. reflexive or automatic response transfers to an unrelated stimulus/behaviour.
2. the generic name for the empirical concepts defined procedurally below. The construction of associations or conditioned reflexes. Dur-

ing conditioning the unconditioned stimulus is replaced by the conditioned stimulus as the factor that releases the given response.

3. the hypothetical production of a change in the nervous system presumed to occur with and to underlie conditioning.

See also **Operant Conditioning, Classical Conditioning.**

- **conflict (motivational)**

1. a term applied when the stimuli for two incompatible responses are presented simultaneously under conditions in which either, presented alone, would yield a response.

2. the state of the animal when two drives associated with incompatible behaviours are equally or nearly equally strong. Conflict behaviour may result in a number of responses including alternation, redirected and displacement behaviour.

3. denotes a state of motivation in which tendencies to perform more than one activity are simultaneously expressed. At any particular moment the animal has many incipient tendencies but by a process of decision making one of these becomes dominant, i.e., the consummatory behaviour.

- **congeneric**

species belonging to the same genus.

- **conifer**

a tree that produces cones, such as a pine, spruce or fir tree.

- **connective**

relatively large nerve trunk in invertebrates consisting of axons that connect one ganglion of the central nervous system to another.

- **connectivity (of habitats)**

the linkage of similar but separated vegetation stands by patches, corridors or 'stepping stones' of like vegetation. This term can also refer to the degree to which similar habitats are linked.

- **consanguinity**

two or more individuals are said to be consanguineous if they have a common recent ancestor (usually not further back than three or four generations.)

- **consciousness**
immediate awareness of things, events and relations.

- **conservation**
the principles and practice of the science of preventing species extinctions.

- **conservation biology**
multidisciplinary science created to deal with the crisis of maintaining the genes, species, communities, and ecosystems that make up earth's biological diversity. Its goals are to investigate human impacts on biodiversity and to develop practical approaches to preserving biodiversity and ecological integrity.

- **conservation-tillage farming**
crop cultivation in which the soil is disturbed little (minimum-tillage farming) or not at all (no-till farming) to reduce soil erosion, lower labour costs and save energy.

- **conservative**
a character that resists or does not change over evolutionary time.

- **consort pair**
in primates a temporary male-female partnership in a polygynous group during the female's estrous.

- **conspecific**
1. belonging to the same species.
2. considered by taxonomists to be in the same species.

- **conspicuous character**
certain anatomical characters are presented during a display which have a particularly obvious aspect.

- **consumer surplus**
the difference between the price actually paid for a good, and the maximum amount that an individual is willing to pay for it. Thus, if a person is willing to pay up to Rs. 3 for something, but the market price is Rs. 1, then the consumer surplus for that item is Rs. 2. This measure approximates, and is bounded by, the more technically precise measures of economic benefit: compensating variation or equivalent variation.

- **consummatory act**
the last phase of an appetitive behaviour sequence which ends the elicited series of actions and leads under

normal conditions to drive reduction. A response that most often terminates a given, frequently occurring sequence of behaviour patterns.

- **consumption efficiency**
the percentage of energy available that is actually consumed at a trophic level. In the case of herbivores it is the percentage of net primary productivity that is ingested.

- **consumptive use**
use of resources that reduces the supply, such as logging and mining.

- **contest competition**
intraspecific competition in which mortality compensates exactly for increases in density, so that there are a constant (or approximately constant) number of survivors irrespective of initial density.

- **context**
the situation in which a signal occurs, which can influence the effect of the signal on a recipient. For example, the song of a male songbird will attract an unmated female in a courtship context but will repel a rival male in a context of territorial defence. The context includes everything besides the signal influencing the effect on a particular occasion. It can include, as in the example given the sex of the recipient and the other signals given in sequence or in concert with the signal in question.

- **continental drift**
1. the separation and movement of land masses in geological time.
2. the movement of the continents with respect to one another.

- **continental shelf**
that part of the continental margin between the shoreline and the continental slope.

- **continentality**
climatic conditions associated with locations in the middle of large continents.

- **contingency**
1. spatial or temporal association. The joining (simultaneous or sequential) occurrence of two behaviour patterns, or a stimulus and a behaviour pattern, or a behaviour pattern and rein-

forcement, such as can give rise to persistent linkage. The contingency of reinforcement plays an important part in operant conditioning; some would say it has an essential, if not defining, role. 2. Also under natural conditions temporal relations between stimulus and response can be of considerable significance, especially in social interactions. Among some birds, the unhitched young learn prenatal to distinguish the calls of their own parents from those of neighbours, apparently because the chicks' parents immediately answer the chicks' calls, and no such contingency occurs for the vocalisations of the neighbouring birds.

- **contour**
  a line drawn on a map connecting points of the same elevation.

- **contrast**
  difference in the relative brightness of two objects.

- **Control Action Threshold (CAT)**
  the combination of pest density and the densities of the pest's natural enemies beyond which it is necessary to intervene and take control measures against the pest, to prevent its population rising to a level at which it will cause economic damage.

- **control of behaviour**
  refers to the underlying mechanisms of behaviour.

- **control theory**
  that part of cybernetics dealing with automatic regulation of machines, including organisms. It is highly general in application, hence abstract and mathematical. Its influence in ethology is most prominent in theorising and research feedback mechanisms in animal behaviour.

- **control**
  a parallel experiment, test or treatment carried out to provide a standard against which an experimental result can be compared.

- **convection cell**
  in plate tectonics, a pattern of mass movement of mantle material in which the central area is uprising and the outer area down flowing, due to heat variations.

- **convergence**
  a superficial resemblance between two organisms or organs that has evolved under the influence of similar selective pressures, but from widely different ancestries.

- **convergent evolution**
  the process by which organisms of different evolutionary lineages come to have similar form or behaviour.

- **coolidge effect**
  the situation in which a sexually satiated male will again begin to court and copulate if presented with an unfamiliar female with which he has not recently mated.

- **cooperative behaviour**
  mutualistic behaviour that benefits both parties and which normally involves complex motor acts and reciprocity.

- **cooperative breeding**
  1. the term cooperative breeding was used as a synonym for communal breeding by a few authors in the early 70s.
  2. it may be used more restrictively to refer to the situation in which two or more breeders cooperate in rearing joint young.

- **coordination**
  muscular movements must be coordinated for behaviour to occur. There are two main processes controlling coordination. One is central control where there is a precise series of instructions that are issued by the brain and obeyed by the muscles. Alternatively coordination comes through peripheral control and is achieved through sense organs in the muscles which send information to the controlling mechanism in the ganglion and thereby influences the instructions issued from the brain.

- **cope's rule**
  the common phylogenetic trend toward increasing size within a lineage.

- **coprophagia**
  refers to eating feces.

- **coprophagy**
  the consumption of feces.

- **copulation**
  a part of sexual behaviour which is most closely asso-

ciated with fertilisation of the egg by sperm.

- **coral bleaching**
  the loss of colour from a coral as it expels its zooxanthellae- usually a stress response.

- **core area**
  area of heaviest use, may centre on a resource, i.e. nest, water source, food source.

- **core**
  central zone of the earth, divided into two zones, the liquid outer core (2900-5080 km deep) and the solid inner core (5080 to centre).

- **correlation**
  a statistical relationship that quantifies the degree to which two variables are associated.

- **correlogram**
  a statistical procedure for the analysis of time-series data and to test for the presence of cyclic phenomena.

- **corridor**
  elements of the landscape that connect similar areas. Streamside vegetation may create a corridor of willows and hardwoods between meadows where wildlife feed.

- **cost of meiosis**
  the cost (in fitness) associated with sexual reproduction where half the reproduction goes into the production of males.

- **cost**
  a decrement in something of value to an animal, e.g. energy. Costs are weighed against benefits.

- **cost-benefit analysis**
  estimates and comparison of short-term and long-term costs (losses) and benefits (gains) from an economic decision. If the estimated benefits exceed the estimated costs, the decision to buy an economic good or provide a public good is considered worthwhile.

- **coupled oscillations**
  linked fluctuation in the abundance of two species, broadly speaking a 'predator' and a 'prey', in which low prey abundance leads to

low predator abundance which leads to high prey abundance, which leads to high predator abundance which leads to low prey abundance and so on.

- **coupling**
  of a gamete or chromosome, bearing at two or more loci alleles that have been designated to be alike in some way (e.g., both wild type rather than mutant.)

- **courtship behaviour**
  behaviour patterns and signals that attract the sex partner and lead to mating and conception.

- **courtship colouration**
  the colour patterns occurring solely in connection with courtship, the mating season or behaviour related to courtship.

- **courtship feeding**
  an epigamic behaviour pattern in which the members of a pair feed one another or where one individual (usually the male) feed the others.

- **courtship**
  period of pair formation and precopulatory behaviour.

- **covariance**
  a statistical term for a component of the correlation coefficient.

- **cover**
  any feature that conceals wildlife or fish. Cover may be dead or live vegetation, boulders, or undercut streambanks. Animals use cover to escape from predators, rest, or feed.

- **cover forage ratio**
  the ratio of hiding cover to foraging areas for wildlife species.

- **cover type (forest cover type)**
  stands of a particular vegetation type that are composed of similar species. The aspen cover type contains plants distinct from the pinyon-juniper cover type.

- **coy**
  the behaviour of one individual toward the sexual displays of another in which the individual does not cooperate or appears not to be interested although sexually receptive.

- **Crassulacean Acid Metabolism (CAM)**
  a pathway for the assimilation of carbon dioxide by plants, in which carbon dioxide is fixed into organic acids during the night and released during the day for photosynthesis. Characteristic of succulent desert plants.

- **created opening**
  an opening in the forest cover created by the application of even-aged silvicultural practices.

- **crèche**
  a group of the young of several families tended by one or more adults.

- **crèching**
  leaving one's young in a social group attended by conspecific adults whiie out foraging, as in penguins.

- **crepuscular**
  pertaining to activity at dawn and dusk.

- **cretaceous period**
  a geological era; the final part of the Mesozoic, from approximately 140 to 170 million years ago.

- **critical habitat**
  areas designated for the survival and recovery of federally listed threatened or endangered species.

- **critical period**
  a phase in an animal's life when certain experiences are particularly likely to have a potent developmental effect.

- **cronism**
  1. a type of cannibalism in which the parents eat their young.
  2. a form of cannibalism. Refers to parents eating their own young. If young are killed but not eaten the appropriate term is infanticide.

- **cross fostering**
  an experimental design in which young of one parent are reared by another. See also **infostering**.

- **cross gait**
  alternating step in which the front and rear leg of each side move in opposition to each others progression in anterior limbs occurs when the corresponding rear leg is on the ground.

• **crossbreeding**
sexual reproduction in which gametes originate form animals belonging to different breeds.

• **crossing-over**
the process of exchange of genetic information between two homologous chromosomes, presumed to occur through breakage of both chromosomes at homologous sites followed by reunion after exchange.

• **crowding**
an unusually high spatial density of animals which may cause discomfort or stress.

• **crowing**
a type of vocalisation produced by male chickens which has frequency oscillations of wide amplitude, distinct breaks and a duration of about 2 sec. Thought to have a territorial or mate attracting function.

• **crown height**
the distance from the ground to the base of the crown of a tree.

• **crust**
the outermost shell of earth extends to a depth of 30-50 km below continents and 10-12 km below oceans.

• **crypsis**
form or behaviour of an organism that makes it difficult to detect, e.g. camouflage from a predator.

• **cryptic coloration**
concealing, or coloration that makes an animal resemble its background.

• **cryptic**
refers to a colour pattern or behaviour that allows the animal to remain hidden against the background.

• **cuckoldry**
refers to the situation in which a male is helping to rear young that are not related to him.

• **cuckoo behaviour**
See **brood parasitism**.

• **cultural behaviour**
the passing of information from one generation to the next by non-genetic means.

• **cultural evolution**
changes in traits of organisms due to information transferred through cultural rather than genetic means.

- **cultural resource**
  the remains of sites, structures, or objects used by people in the past; this can be historical or pre-historic.

- **culture**
  a body of nongenetic information transmitted from generation to generation or within generations; the transfer of information by behavioural means.

- **cumulative effects**
  effects on the environment that result from separate, individual actions that, collectively, become significant over time.

- **curiosity**
  behaviour associated with the examination and exploration of new objects, areas or situations.

- **current**
  moving charge in an electrical circuit; Unit: ampere.

- **cursorial**
  adapted for running.

- **cuticle**
  in plants, a layer of waxy substance on the outer surface of epidermal cell walls. In animals, the outermost layer of many invertebrates.

- **cyanogenic**
  releasing hydrogen cyanide.

- **cyclic reflex**
  an alternative hypothesis to the central pattern generator to explain rhythmic behaviour such as walking or swimming. This hypothesis posits feedback control of the behaviour whereas the CPG hypothesis open-loop control.

- **cytochrome**
  a group of iron-containing hydrogen or electron carriers in cellular metabolism.

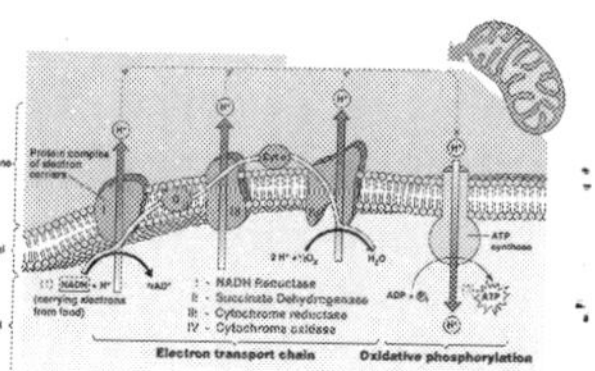

- **cytoplasm**
  the living matter within a cell, excluding the genetic material.

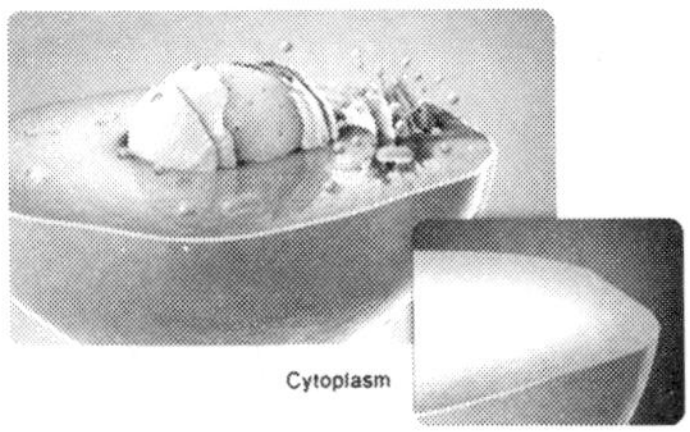

- **d**
  the individual death rate, the average number of deaths per individual per unit time.

- **dabbling**
  a feeding behaviour of anatid ducks, in which the head is thrust below the water and the posterior end of the duck is raised vertically out of the water.

- **dance language**
  a general term for communication among bees with regard to the distance and direction of a food source, nest site or other goal.

- **dark adaptation**
  adaptation to decreasing levels of ambient illumination by an increase in receptor sensitivity.

- **Darwin**
  a unit of measure, the doubling of size in one million years is one Darwin.

- **DBH (Diameter at Breast Height)**
  the diameter of a tree 4 and 1/2 feet above the ground on the uphill side of the tree.

- **deafferentation**
  the severing of all afferent nerves, for example, the dorsal roots of the spinal nerves. This is done to determine the endogenous activity of a system.

- **death feigning**
  a form of defensive behaviour in which the animal becomes immobile as if dead. Also called thanotosis.

- **death shake**
  many carnivores especially the candies pick up the prey loosely with the teeth and shake it vigorously and flip it aside. This shaking can result in respiratory collapse in the prey. More intense shaking leads to death by dislocating the neck.

- **debt-for-nature swap**
  agreement in which a certain amount of foreign debt is cancelled in exchange for local currency investments that will improve natural resource management or protect certain areas in the debtor country from harmful development.

- **deception**
  the concealment or misrepresentation of truth which leads another (individual) into error or to disadvantage. In humans this clearly im-

plies intent but among animals intent is neither known nor necessary.

- **decision**
an animal may perform alternative actions at various points during a sequence of behaviour. A decision must have been made at each of these choice points. (does not imply cognition). A commitment of time to a course of action.

- **decision criteria**
the rules and standards used to evaluate alternatives to a proposed action on National Forest land. Decision criteria are designed to help a decision maker identify a preferred choice from the array of alternatives.

- **decision rule**
the mechanism controlling a decision.

- **decking area**
a site where logs are collected after they are cut and before they are taken to the landing area where they are loaded for transport.

- **decomposition**
the breakdown of complex, energy-rich organic molecules to simple inorganic constituents.

- **dedicated circuit**
neural configuration of particular sensory, inter- and motor neurons responsible for generating a specific behaviour.

- **defecation**
elimination of faeces from the body.

- **defensive behaviour**
refers to any behaviour which reduces the chances of one animal being harmed by another.

- **defensive threat**
a type of threat which occurs while the animal is escaping a predator.

- **deforestation**
1. the large-scale removal of trees from a habitat dominated by forest.
2. removal of trees from a forested area without adequate replanting.

- **degradative succession**
  a temporal succession of species that occurs on a degradable resource.

- **DEIS (Draft Environmental Impact Statement)**
  the draft version of the Environmental Impact Statement that is released to the public and other agencies for review and comment.

- **deletion**
  the loss of a segment of chromosome.

- **demand curve**
  1. the graphical representation of the demand function. The demand function relates price and quantity demanded. It tells how many units of a good will be purchased at different prices.
  2. In general, at higher prices, less will be purchased, so demand curves slope downward. The market demand function is calculated by adding up all of the individual consumers' demand functions.

- **demand function**
  the mathematical function that relates price and quantity demanded for goods or services. It tells how many units of a good will be purchased at different prices. The market demand function is calculated by adding together all of the individual consumers' demand functions.

- **deme**
  an interbreeding local population of a given species.

- **demic selection (or inter-demic selection)**
  1. the disproportionate growth of a subset of a population or species. Demic selection has an effect on the general genetic composition of the population if the subsets have different gene frequencies.
  2. demic selection is a type of intergroup selection.

- **demographic process**
  a process capable of changing the size of a population, viz. birth, death or migration.

- **demographic transition**
  hypothesis that countries, as they become industrialised, have declines in death rates followed by declines in birth rates.

- **demography**
  the study of populations and how they survive, die, reproduce and grow.

- **dendrite**
  1. the short processes of a neuron that carry information to the cell.
  2. one of the branching small neural processes carrying signals to the cell body of a neuron.

- **density dependence**
  1. the tendency for the death rate in a population to increase, or the birth or growth rate to decrease, as the density of the population increases.
  2. effects correlate with population density.
  3. the tendency for the death, birth or growth rate in a population neither to rise nor fall as the density increases.

- **density-dependent selection**
  selection in which the fitness of a genotype depends upon the density of the population in which it occurs. As opposed to density independent.

- **Deoxyribo Nucleic Acid (DNA)**
  the carrier of genetic information in cells; capable of self-replication as well as coding for RNA synthesis.

- **dependent variable**
  in an experiment, the effect that is measured when the independent variable is varied.

- **depolarisation**
  change in the cell's membrane potential where the interior becomes less negative (i.e., a shift towards zero.)

- **depression**
  in a behavioural sense, a state of severe dejection characterised by reduced activity, seclusion and or stereotyped actions often directed to the body.

- **deprivation experiment**
  see **isolation experiment**.

- **derived**
  a characteristic that has originated within a taxon.

- **descending contralateral movement detector**
  neuron which in locusts and grasshoppers responds vigorously to images of approaching objects; the axon crosses the brain descending from hindbrain to thorax to excite motor neurons concerned with jumping and flying and responds to movements detected by the opposite eye.

- **desensitisation**
decrease of a response to a given stimulus with practice or with repeated exposure. The process usually begins by presenting the stimulus at a reduced level so that the animal does not respond (e.g., a recording of thunder at reduced volume) and gradually increasing the intensity on repeated exposures until the animal responds.

- **desert**
a desolate and barren region, usually deficient in available water and with scant vegetation.

- **desertification**
conversion of rangeland, rain-fed cropland to desert-like land, with a drop in agricultural productivity of 10% or more. It is usually caused by a combination of overgrasing, soil erosion, prolonged drought, and climate change.

- **desired future condition**
land or resource conditions that are expected to result if goals and objectives are fully achieved.

- **despotic**
refers to the social situations in which one individual controls the activities of others.

- **deterministic forces**
forces which can be characterised exactly with no element of chance or probability.

- **detour experiment**
an experiment to evaluate spontaneous insight behav-iour or the ability of the animal to figure out how to locate an object that it can reach only by a circuitous route. The solution to the detour problem is usually rewarded with food.

- **detritivory**
consumption of dead organic matter (detritus) usually together with associated microorganisms.

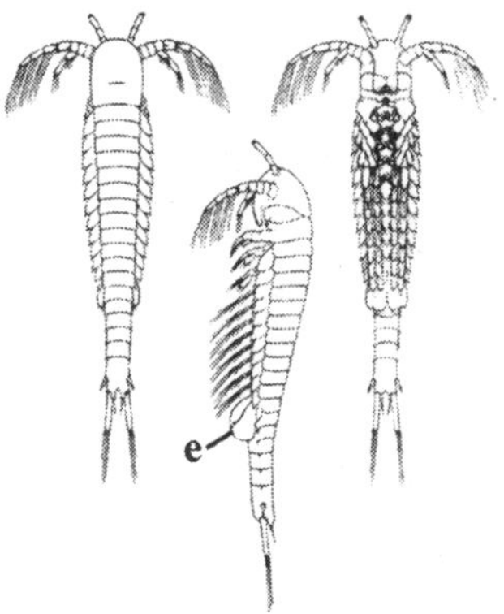

- **developed recreation**
  recreation that requires facilities that, in turn, result in concentrated use of the area. For example, skiiing requires ski lifts, parking lots, buildings and roads. Campgrounds require roads, picnic tables and toilet facilities.

- **development**
  synonymous with ontogeny. It is concerned with the history of the individual from conception to death and with changes in behaviour that are associated with different life stages.

- **development control plan**
  a planning instrument that can cover a wide range of issues, but with more flexibility and usually in more detail than a local plan. It may include planning guidelines for particular areas or types of development.

- **developmental homeostasis**
  the capacity of developmental pathways to produce a normal phenotype in spite of developmental or environmental disturbance.

- **developmental threshold**
  the body temperature of an organism below which no development occurs.

- **devonian period**
  a geological era lasting from approximately 400 to 360 million years ago.

- **dialect**
  as defined by ethologists refers to the vocalisations (or other signal) of a population of animals that differ from those of another population of the same species.

- **diapause**
  a state of arrested development or growth, accompanied by greatly decreased metabolism, often correlated with the seasons, usually applied only to insects.

- **diapause**

  the period in an insect life cycle when no reproduction takes place.

- **dicotyledon**

  a member of one of the two classes of flowering plants, distinguished by having seedlings with usually a pair of seedling leaves (cotyledons) and commonly with floral parts in fours or fives, leaves with net venation and ability to form wood by secondary (cambial) cell division within the tissues.

- **differential resource utilisation**

  normally used only in the context of interspecific competition, the use of different resources by two different species, or the use of the same resource at a different time, in a different place, or generally in a different manner.

- **differentiation**

  the development of organs and body parts in ontogeny from simpler antecedent structures.

- **diffraction**

  waves are scattered when they pass through an aperture or are bent by passing from one medium to another.

- **diffusion coefficient**

  a measure of the rate of movement of solutes or gases in response to a concentration gradient in the medium in which they are dissolved.

- **dilution effect**

  safety in numbers; the decreased probability of being taken by a predator that stems from being a member of a large group of prey as opposed to a small group.

- **dimorphism**

  1. the existence of two distinct forms of an organism or organ, e.g. winged and wingless generations in the life of aphids, winged and wingless seeds produced from the same flower.

  2. occurrence of two distinct morphological types in a single population.

- **dioecious**
  refers to the situation in which the sexes occur in separate individuals.

- **dioxin**
  a synthetic, organic chemical of the chlorinated hydrocarbon class. It is one of the most toxic compounds known to humans, having many harmful effects, including induction of cancer and birth defects, even in extremely minute concentrations. It has become a widespread environmental pollutant because of the use of certain herbicides that contain dioxin as a contaminant.

- **diploid**
  a cell or organism having a double set of chromosomes; the normal chromosome number in the cells (except mature germ cells) in any individual derived from a fertilized egg. See **polyploid**.

- **direct fitness**
  the long-term reproductive success of an individual through its direct offspring. Inclusive fitness minus indirect fitness. Direct fitness is synonymous with classical fitness and with individual fitness. Direct fitness has a personal component and a kinship component. The latter is responsible for the evolution of parental care.

- **direct fitness**
  probability of reproductive success through one's own offspring.

- **direct head scratching**
  scratching pattern in birds in which the claw is brought up directly to the head under the wing. See **indirect head scratching**.

- **direct selection**
  1. natural selection (including sexual selection) based on direct fitness. See **direct fitness.**
  2. selection resulting in a shift in the population mean for the character considered.

- **directional selectivity**
  most vigorous neuronal response to stimulation from a particular direction.

- **disassortative mating**
  a pattern of mating in which unlike phenotypes mate more frequently than expected when mated at random.

• **disaster**

major disturbances in the life of a community or population which occur sufficiently often to leave their record in the 'genetic memory' of the population.

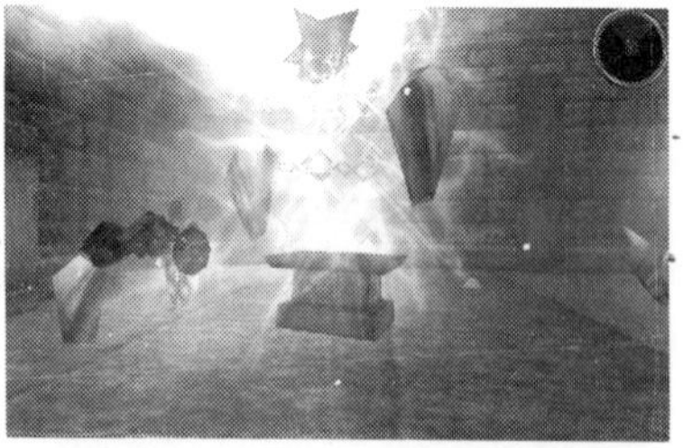

• **discharge**

with reference to a drive, the utilisation, consumption, annulment, or destruction of a drive (or of an accumulation of action-specific energy or of 'impulses') that occur when its consummatory act is given.

• **discount rate**

the rate used to reduce future benefits and costs to their present time equivalent.

• **discrete generations**

a series of generations in which, strictly, each one finishes before the next begins. Commonly, however, the early life cycle stages of the succeeding generation overlap with the end of the final stages of the preceding generation.

• **discrimination**

1. differential response to two or more stimuli.

2. ahypothetical sensory or neurophysiological event which makes it possible for an animal to form discriminations as defined above. A capacity.

• **disease**

the disturbed or altered condition of an organism (malfunctioning) caused by the presence of an antagonist (toxin or pathogen) or the absence of some essential (e.g. micronutrient or vitamin.)

• **disgust reaction**

contraction of the pharyngeal musculature, salivation and behaviour resembling vomitting.

- **disinhibition**

1. an explanation for displacement behaviour. This is the notion that many actions are competing for access to the behavioural final common path. When all others are blcɔked, then a low priority activity like grooming may be distributed.
2. the term applied to the observation, in the course of extinction of a classical conditioned response, that, when an extraneous stimulus (one not previously present in the situation and which does not evoke the same response) is presented together with the conditioned stimulus, the magnitude of CR may be larger than predicted from its magnitude on previous trials.

- **disjunct distribution**

the geographical distribution of a species or other taxonomic class of which parts are widely separated.

- **dispersal**

1. the spreading of individuals away from each other, e.g. of offspring from their parents and from regions of high density to regions of lower density.
2. travel of Individuals, which as an ecological process affects distribution of individuals and as a genetic process affecting geographic differentiation and variation.

- **dispersal polymorphism**

two or more types of dispersal structures found within a species or among the progeny of an individual.

- **dispersed recreation**

recreation that does not occur in a developed recreation site, such as hunting, backpacking, and scenic driving.

- **dispersion**

refers to the distribution in space of individuals within their natural habitat.

- **displacement activity**

1. traditionally regarded as an irrelevant activity performed due to excess energy which cannot be released through normal channels. Most frequently this is caused by conflicts between opposing drives. Subsequently displace-

ment behaviour was used to describe any irrelevant pattern without any implications as to why it might occur. Today the term is not used due to a number of conceptual difficulties.
2. in psychology displacement refers to what ethologists called redirected behaviour. A response that has been identified as a member of one instinct and often occurs in others. On these occasions it is termed a displacement activity.
3. a displacement activity is an activity belonging to the executive motor pattern of an instinct other than the instinct(s) activated.
4. a term applied to a response that has been conditioned under one set of deprivations and reinforcement conditions and that then appears under others.
5. behaviour patterns seemingly unrelated to the behavioural context in which it occurs.

- **display behaviour**
behaviour which has evolved in its role as a communication signal, i.e. an adaptation for its role in communications.

- **display drinking**
1. epigamic behaviour pattern in many ducks, in which the male dips the bill in the water and raises it as though he was drinking but he does not. Display behaviour in many species of Anatid ducks. The individual dips its bill in the water as though it were about to drink, but no water or swallowing occurs. This behaviour is more stereotyped than is normal drinking behaviour.
2. ritualised drinking behaviour.

- **display flight**
in song birds, especially species of open areas, the birds make conspicuous, stereotyped flights which serve to attract females and to indicate territorial possession to conspecifics.

- **display or sham preening**
display behaviour in many species of anatid ducks. Drakes raise the wing as if they were about to preen, but the flight feathers are only lightly touched as if to draw attention to the colourful

speculum. The behaviour is more stereotyped than normal preening.

- **disruptive selection**
  selection in which two or more different genotypes are at an advantage and the intermediate types are at a disadvantage. (Same as diversifying selection).

- **distraction display**
  behaviour pattern serving to divert a potential predator from the nest; common in ground-nesting birds. See **broken-wing display**.

- **distress calls**
  vocalisation which serves to bring the separated young animal once again into contact with its parent or social companions.

- **distribution**
  1. the spatial range of a species, usually on a geographic but sometimes on a smaller scale, or the arrangement or spatial pattern of a species over its habitat.
  2. systematic arrangement of numerical data according to some ranked measurement scale.

- **disturbance**
  1. in community ecology, an event that removes organisms and opens up space which can be colonised by individuals of the same or different species.
  2. any event, such as forest fire or insect infestations that alter the structure, composition, or functions of an ecosystem.

- **dithering**
  to act irresolutely, to vacillate; unable to decide among options.

- **diuresis**
  increased formation of urine.

- **diuretic**
  any agent that increases formation of urine.

- **diurnal**
  relating to activity during the day.

- **divergence**
  adaptations that occur through alterations of the ancestral types.

- **divergent evolution**
  the evolution by natural selection of differences among

closely related species that live in different environments.

- **diversity**
  see **Species diversity**.
- **diversity**
  of species or other taxa, usually the number present in a particular ecological unit .May also include a measure of the distribution of population sizes among species.
- **diversity index**
  a mathematical index of species diversity in a community.
- **division of labour**
  in social insects there is a division of the different activities of the colony among different castes. This often refers to a reproductive division of labour, where one or a few individuals lay all the eggs or get all the matings whereas most are workers.
- **dizygotic twins**
  twins that develop from two zygotes, no more related than other siblings.
- **DNA**
  Deoxyribo-Nucleic Acid. Giant molecules that in most organisms are the nuclear repository for the genetic information and which are replicated and transmitted equally to daughter nuclei.

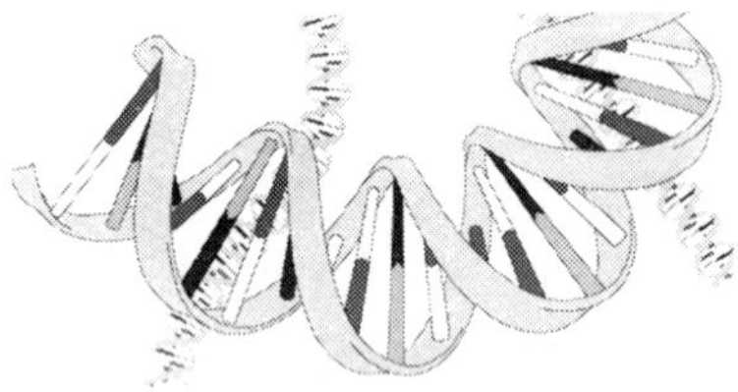

- **dog-sitting position**
  resting on the caudal part of the body observed in pigs, cattle and horses and often indicative of disease.
- **domestication**
  the process by which man has modified certain species of animal by maintaining them in or near human habitations and by artificial selection.
- **dominance**
  this term has specific meanings in three branches of biology.
  1. in genetics dominance exists between two alleles when the dominant one commands phenotypic expression in the heterozygous condition; the other allele, termed recessive, is pheno-

typically expressed only in the homozygous condition.
2. in ecology dominance refers to the relative abundance of a species in comparison with other species in a biological community.
3. in ethology dominance refers to superior position in a rank order or social hierarchy. An individual to which another consistently gives way is said to be dominant in the relationship, the other being subordinate or subdominant.

In a linear hierarchy each individual is dominant over those below it and subordinate to those above it, with the exceptions of the alpha animal, who is subordinate to none, and the omega animal, who is dominant over none. However, social hierarchies may be more complicated, as when animal A dominates B, B dominates C, and C dominates A. Thus dominance is relative to individual relationship. Further complications arise if two or more individuals from an alliance and so collectively are dominant over those that could dominate each of them singly. Such arrangements are common in primate groups, where social organization may also include subdivision into classes according to sex and age status, dominated by a central hierarchy of the older males. Dominance status is anything but static in the course of group life, as lower-ranking animals continually test the ability of those above them to hold their place, and as the struggle of higher-ranking animals to maintain position becomes more difficult with advancing age. Dominance ranking may very with context, so who gives way to whom may not be the same in contests over spatial claims, food and sexual access. There are thus problems of definition and assessment of dominance in social relationships as studied by ethologists and sociobiologists.

- **dominance behaviour**
demonstration of superiority in pursuing or maintain-

ing the dominant position within a social group; priority of access to resources.

- **dominance hierarchy**
  a social ranking within a group in which some individuals give way to others, often conceding useful resources to others without a fight.

- **dominance portion of genetic variance**
  that portion of genetic variance for a given trait that results from the fact that heterozygotes do not always score halfway between homozygotes.

- **dominant species**
  species which make up a large proportion of community biomass or numbers.

- **donor-controlled models**
  mathematical models of predator - prey interactions in which the donor (prey) controls the density of the recipient (predator) but not the reverse.

- **dormancy**
  an extended period of suspended or greatly reduced activity, e.g. aestivation and hibernation.

- **dorsal light reaction**
  orientation so that light is kept dorsal to and perpendicular to both long and transverse axes of the body.

- **dosage compensation**
  the mechanism by which genes on the x chromosome have the same physiological effect in females (with xx) as in males (with xy).

- **dreaming**
  1. during human sleep there are phases during which there are rapid eye movements and elevated breathing and heartbeat rates. Associated with this are characteristic patterns of brain activity which resembles the activity of an awake brain. However, the individual is in a deep sleep and not easily aroused. If the person is aroused from this sleep they usually report that they

have been dreaming whereas when awakened from quiet sleep they do not make such reports.

2. many sleeping animals show similar eye movements and brain activity although we do not know whether they have the cognitive experience of dreaming.

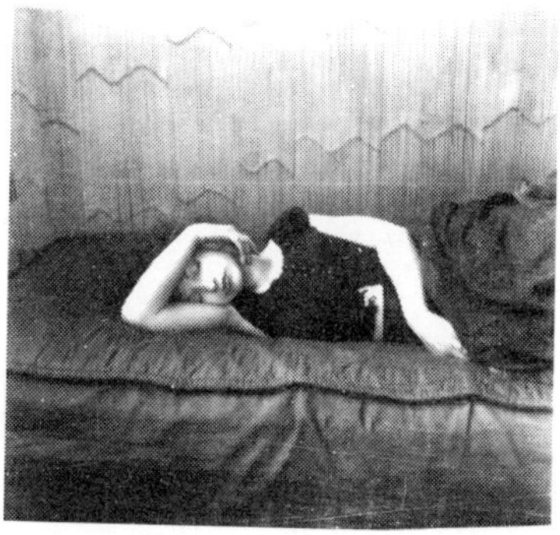

- **drinking**

  the process of ingesting water.

- **drive**

  1. a drive is a specific internal source of energy which is produced in an automatic rhythmic manner and which leads to a state in which the animal engages in a spontaneous behaviour since the drive is pressing for its own release even in the absence of appropriate external stimuli. After the consummatory act is performed, the energy to perform the activity drops and the animal is less likely to perform the action in the immediate future.

  2. a hypothetical state of the animal which is identified by: (a) gross changes in the relative frequency of broad classes of behaviour that are not attributable to disease, learning, or growth; (b) changes in running-wheel activity; or (c) changes in cage-motility. Drives may be manipulated by: operations of deprivation (as of food and water), alterations of the hormone or other biochemical balance of the blood (e.g., of ACTH, testosterone, sodium chloride), by temperature changes, or intense stimulation. Some can be specified only by stating the time of the year.

3. a state of the animal established by deprivation of food or water, or by the presentation of electrical shock, characterised by a change in the relative rates of occurrence of a specified set of operant responses.

- **drumming**

the production of low frequency sounds by various means.

1. intraspecific communication among woodpeckers in which various pecking patterns are used as signals.

2. a common form of communication in insects in the production of low frequency sounds by stridulating, such as the sounds made by queen honeybees.

3. in many rodents the forepaws are beaten on the ground during threat display. Other species may mark the ground by drumming the hind paws after rubbing them across their scent glands.

4. the sound made by some species of grouse from display perches on logs.

- **duetting or duet singing**

a behaviour associated with pair formation and bonding. Partners coordinate the rhythm and types of call of their respective song parts.

- **duplication**

the presence of a section of a chromosome more than once in a genome.

- **dust-bathing**

a behaviour pattern that is a component of integumentary care characterised by lying on the side and making a small depression in the substrate, wing-shaking, bill-raking and scratching.

- **dynamic equilibrium**

the state of a system when it remains unchanged because two opposing forces are proceeding at the same rate.

- **dynamically fragile**
  describes a community which is stable only within a narrow range of environmental conditions.

- **dynamically robust**
  describes a community which is stable within a wide range of environmental conditions.

- **dystocia**
  difficult parturition.

- **e**
  see extinction rate.

- **early forest succession**
  the biotic (or life) community that develops immediately following the removal or destruction of vegetation in an area. For instance, grasses may be the first plants to grow in an area that was burned.

- **ecdysis**
  the act of shedding or molting the outer exoskeleton or cuticle.

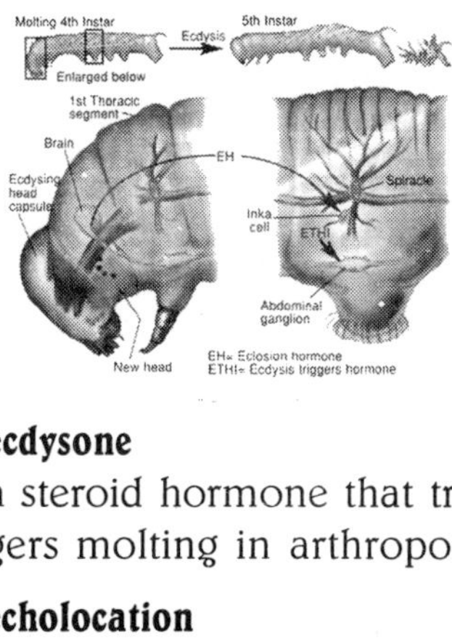

- **ecdysone**
  a steroid hormone that triggers molting in arthropods.

- **echolocation**
  1. the production of ultrasonic sounds which are reflected off objects in the environment and are used for locating the position of objects.
  2. emission of loud pulses of sound and analysis of returning echoes in order to find what lies ahead; used by

many bats for hunting and navigating.

3. the sonar-like ability used by bats, dolphins, and other animals to detect objects. Using echolocation, the animal emits high-pitched sounds that reflect off of an object and return to the ears or other sensory receptors.

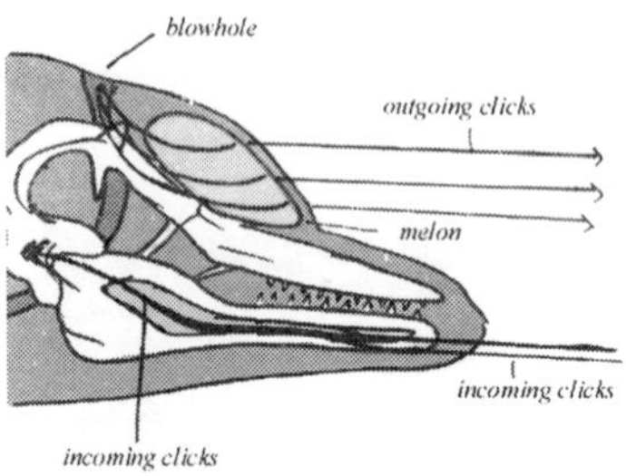

- **eclosion**

emergence of the adult from the pupa.

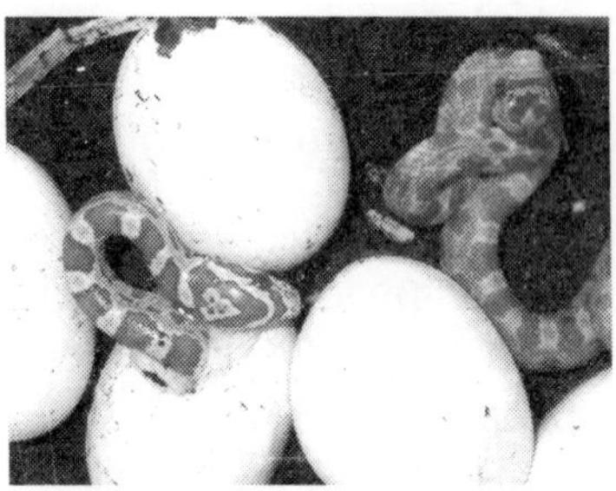

- **ecogeographical rules**

any of the rules describing regularities in the geographical distribution of phenotypic variations (in pigmentation, body size, etc.)

- **ecological approach**

an approach to natural resource management that considers the relationships among all organisms, including humans and their environment.

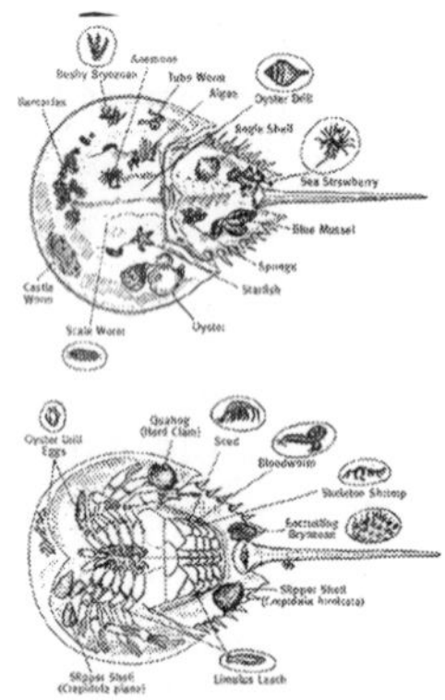

- **ecological efficiency**

1. the ratio of net productivity at one trophic level to net productivity at the next lower level.

2. the percentage of energy in biomass produced by one trophic level that is incorporated into biomass by the next highest trophic level.

- **ecological energetics**

the branch of ecology in which communities are studied from the point of view of the energy flowing through them.

- **ecological fitness**
  the number of a parent's young that live to reproduce; divided by two if sexual reproduction is involved.

- **ecological niche**
  a term with alternative definitions, not all of them synonymous.
  1. the 'occupation' or 'profession' of an organism or species.
  2. the range of conditions, resource levels and densities of other species allowing the survival, growth and reproduction of an organism or species. Hence, if each condition, resource or other species is seen as a dimension, the niche is an *n*-dimensional hyper volume.
  3. the range of combinations of all relevant environmental variables under which a species or population can persist; often more loosely used to describe the role of a species, or the resources it utilises.
  4. the sum total of an organism's utilisation of the biotic and abiotic resources of its environment. The fundamental niche represents the theoretical capabilities and the realised niche represents the actual role.
  5. the sum total of an organism's utilisation of the biotic and abiotic resources of its environment.

- **ecological or ecosystem services**
  valuable services provided by natural systems. Examples of ecological services include flood control, air purification and climate control.

- **ecological processes**
  processes which play an essential part in maintaining ecosystems. Four fundamental ecological processes are the cycling of water, the cycling of nutrients, the flow of energy and biological diversity.

- **ecological pyramid**
  a graphic representation of the quantitative relationships of numbers of organisms, biomass or energy flow between the trophic levels of an ecosystem. Because large amounts of energy and biomass are dissipated at every trophic level, these diagrams nearly always take the form of pyramids.

- **ecological release**
  when a competitor is not present in an area such as on an island and individuals of a species expand their feeding and habitat preferences.

- **ecological species concept**
  the idea that ecological roles (niches) define species.

- **ecological succession**
  1. transition in the species composition of a biological community, often following ecological disturbance of the community; the establishment of a biological community in an area virtually barren of life.
  2. process in which communities of plant and animal species in a particular area are replaced over time by a series of different and often more complex communities.

- **ecologically sustainable development**
  1. ecologically sustainable development is development that uses, conserves and enhances the community's resources so that ecological processes, on which life depends, are maintained and quality of life for both present and future generations is increased. See also **environmentally sustainable development and sustainable development**.
  2. development in which the total human population size and resource use in the world (or in a region) are limited to a level that does not exceed the carrying capacity or the existing natural capital and is therefore sustainable.

- **ecologist**
  a scientist who studies ecology.

- **ecology**
  1. the study of the relationships between living organisms and their environments.
  2. the interrelationships of living things to one another and to their environment, or the study of these interrelationships.
  3. the study of how organisms interact with their environments.
  4. the study of the relationships between organisms and their environments, including: the interactions of living organisms with one another and with their non-

living surroundings, the flow of matter and energy in an environment, and the structure and functions of nature. Also called *bionomics* .

5. the relationship between organisms and their environment.

7. the branch of sociology that is concerned with studying the relationships between human groups and their physical and social environments. Also called human ecology .

8. the study of the detrimental effects of modern civilisation on the environment, with a view toward prevention or reversal through conservation. A component of the field of human ecology.

- **economic efficiency**

the allocation of goods to their highest relative economic value.

- **economic injury level**

the level of pest abundance above which it costs less to control the pest than is saved by pest control, but below which it costs more than is saved.

- **ecophysiology**

the study of physiology and tolerance limits of species that enhances understanding of their distribution in relation to abiotic conditions.

- **ecoregion**

1. an area over which the climate is sufficiently uniform to permit development of similar ecosystems on sites that have similar properties. Ecoregions contain many landscapes with different spatial patterns of ecosystems.

2. a geographically distinct area of land that is characterised by a distinctive climate, ecological features, and plant and animal communities.

- **ecosystem**

1. a holistic concept of the plants, the animals habitually associated with them and all the physical and chemical components of the

immediate environment or habitat which together form a recognisable self-contained entity.
2. biotic community and its abiotic environment; the whole earth can be considered one large ecosystem.
3. a dynamic interdependent complex of living plant, animal, fungal and microorganism communities, the non-living environment they inhabit, and the interactions within and between the living and non-living components.
4. an arrangement of living and non-living things and the forces that move among them. Living things include plants and animals. Non-living parts of ecosystems may be rocks and minerals. Weather and wildfire are two of the forces that act within ecosystems.
5. a community of plants, animals, and microorganisms that are linked by energy and nutrient flows and that interact with each other and with the physical environment. Rain forests, deserts, coral reefs, grasslands, and a rotting log are all examples of ecosystems.
6. the sum of physical features and organisms occurring in a given area.
7. A level of ecological study that includes all the organisms in a given area as well as the abiotic factors with which they interact; a community and its physical environment.
8. an ecological community of various plants, animals, and other organisms, interacting with each other and with the nonliving resources in their environment, all functioning as a unit.

- **ecosystem functions**
the physical, chemical, and biological processes or attributes that contribute to the self-maintenance of the ecosystem; in other words, what the ecosystem does. Some examples of ecosystem functions are wildlife habitat, carbon cycling, or trapping nutrients.

- **ecosystem integrity**
the degree to which the fundamental ecological processes (such as water and nutrient cycling, the flow of

energy and biological diversity) are maintained.

- **ecosystem management**
an ecological approach to natural resource management to assure productive, healthy ecosystems by blending social, economic, physical and biological needs and values.

- **ecosystem services**
1. the beneficial outcomes, for the natural environment, or for people, that result from ecosystem functions. Some examples of ecosystem services are support of the food chain, harvesting of animals or plants, clean water, or scenic views. An ecosystem to provide services to humans, some interaction and some appreciation by, humans is required.
2. services, vital to the support of human life, provided by intact natural ecosystems. These include the purification of air and water, detoxification and decomposition of wastes, regulation of climate, regeneration of soil fertility, and production and maintenance of biodiversity, from which key ingredients of our agricultural, pharmaceutical and industrial enterprises are derived. Historically, the nature and value of Earth's life support systems have largely been ignored until their disruption or loss highlighted their importance.

- **ecotourism**
the enterprises involved in promoting tourism of unusual or interesting ecological sites. Environmentally, culturally, and scientifically responsible tourism that takes great efforts to ensure tourism revenues benefit the local communities where tourism occurs, the local inhabitants benefit the most economically (revenues are not returned to the traveller's country of origin) and native culture is not diluted with imported tourist cultures. Ecotourism safeguards the nature of the attraction that instigated the tourism and serves to strengthen conservation and scientific research efforts in the area. Very few large corporations who claim to engage in ecotourism actually do so.

- **ecotome**
  a transition zone between two diverse communities.

- **ecotone**
  the transition zone between two biotic communities, such as between the Ponderosa pine forest type and the mixed conifer forest, which is found at higher elevations than the pine.

- **ecotype**
  1. a subset of individuals within a species with a characteristic ecology.
  2. a local strain (or sometimes subspecies) whose distinctive features develop under the selective forces in a local environment. A genetically determined phenotype of a species that is found as a local variant associated with certain ecological conditions.
  3. a population of a species in a given ecosystem that is adapted to a particular set of environmental conditions.
  4. a locally adapted variant of a species, differing genetically from other ecotypes of the same species.

- **ectoderm**
  1. the outermost tissue layer of an animal embryo. Also, tissue derived from an embryonic ectoderm.
  2. the outermost of the three primary germ layers in animal embryos; gives rise to the outer covering and, in some phyla, the nervous system, inner ear and lens of the eye.

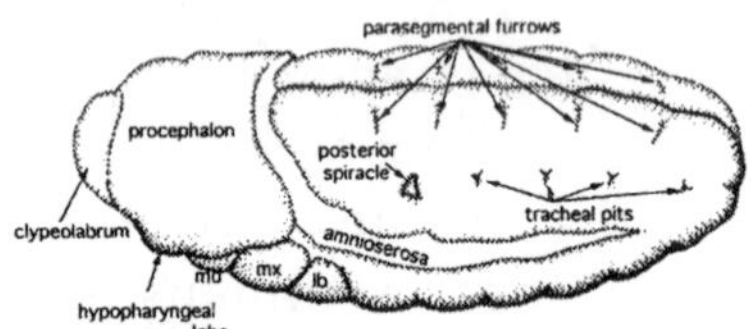

- **ectomycorrhisa**
  an association of a fungus and the root of a plant (usually a tree) in which the fungus forms a sheath around the root and penetrates *between* the cells of the host.

- **ectoparasite**
  a parasite that lives on the surface of its host.

- **ectotherm**

1. an organism in which the body temperature relies on sources of heat outside itself.
2. an animal that approximately corresponds in body temperature to the temperature of the environment.
3. an organism that uses environmental heat and behaviour to regulate its body temperature.
4. an animal such as a reptile, fish, or amphibian, that must use environmental energy and behavioural adaptations to regulate its body temperature.

- **edaphic**

pertains to the soil.

- **edge**

the margin where two or more vegetation patches meet, such as a meadow opening next to a mature forest stand, or a ponderosa pine stand next to an aspen stand.

- **Edward Jenner**

a pioneer of vaccination; used vaccination with material from cowpox lesions to protect people against smallpox.

- **effective population size**

the size of a genetically idealized population with which an actual population can be equated genetically.

- **effector**

1. a general term for muscle or gland that are innervated and thus under control of CNS.
2. the part of an organism that produces a response to a stimulus.

- **efference**

nerve impulses running from the central nervous system to the periphery in contrast to afference. The efferent nerves control the function of the skeletal musculature as well as certain inner organs and effectors.

• **efferent**
carrying away from a centre, applied to nerves and blood vessels.

• **egg**
1. an egg cell or female gamete. Also a structure in which embryonic development takes place, especially in birds, (consists of an egg cell, membranes and often a shell.)

2. a female gamete, which usually contains abundant cytoplasm and yolk; nonmotile and often larger than a male gamete.

• **egg dumping**
the act of a female bird in laying eggs in the nests of another bird.

• **egg-rolling**
motor pattern frequently observed in ground breeding birds by means of which eggs that have rolled outside the nest are retrieved. The graylag goose retrieves an egg with the bill gently rolling it while balancing it with the underside of the bill. If the egg is removed during the procedure, the goose completes the instinctive egg-rolling anyway, but without the orientation component.

• **eggshell removal**
a behaviour in which a parent bird removes eggshells from the nest shortly after the young have hatched. In many species this is a specific action pattern that has been demonstrated to reduce the risk of aerial predation.

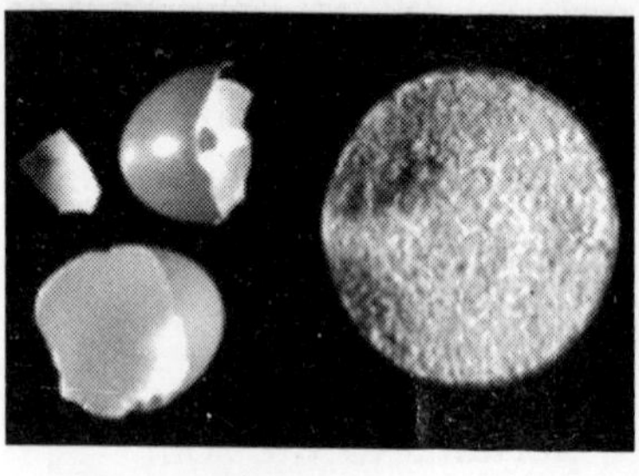

• **ejaculation**
release of sperm during copulation.

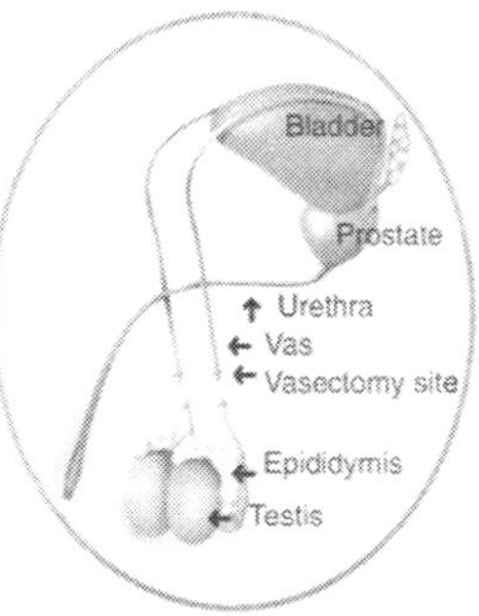

- **ejaculatory duct**
in the male, a duct from each testis that join to form the urethra.

- **el nino southern oscillation**
flip-flopping pressure systems in the South Pacific that trigger short-lived global changes in climate. Warm waters from the western Pacific move across the ocean, just below the equator, and significantly warm the eastern tropical Pacific.

- **electivity**
a measure of the preference, or lack of it, shown by a consumer species for its range of prey.

- **electric potential**
the difference in the amount of electric charge between a region of positive charge and a region of negative charge. The establishment of electric potentials across the plasma membrane and across organelle membranes makes possible a number of phenomena, including the chemiosmotic synthesis of ATP, the conduction of nerve impulses and muscle contraction.

- **electrical synapse**
1. type of synapse in which communication is by electrical synaptic transmission. Potentials are transmitted by direct flow of ions through gap junctions.
2. a junction between two neurons separated only by a gap junction, in which the local currents sparking the action potential pass directly between the cells.

- **electrocardiogram**
a plot of electrical activity of the heart over the cardiac cycle; measured via multiple skin electrodes.

- **electrochemical gradient**
1. combined electrostatic and osmotic-concentration gradient, such as the chemi-

osmotic gradient of mitochondria and chloroplasts.
2. the diffusion gradient of an ion, representing a type of potential energy that accounts for both the concentration difference of the ion across a membrane and its tendency to move relative to the membrane potential.

- **electrocommunication**

the capacity of electric fish to transmit social signals by electric organ discharges.

- **electrogenic pump**

an ion transport protein generating voltage across a membrane.

- **electromagnetic spectrum**

1. the entire spectrum of radiation; ranges in wavelength from less than a nanometer to more than a kilometre.

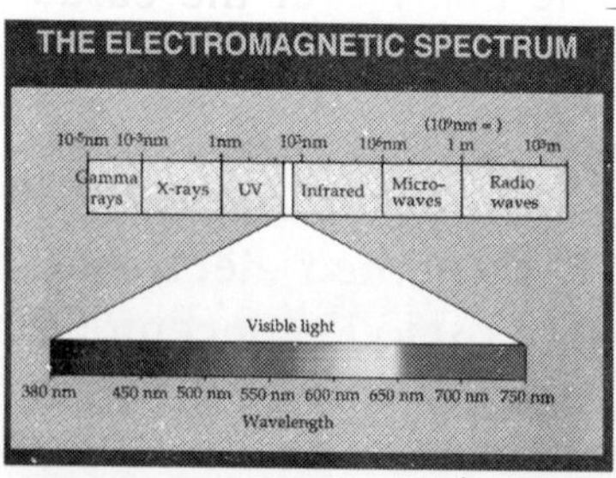

- **electromyograms**

recording of the electrical activity of a muscle by placing extra cellular electrodes on or near the muscle.

- **electron**

1. a negatively charged subatomic particle of an atom or ion. In atoms, the number of electrons present is equal to the number of positively charged protons present. Hence, atoms are electrically neutral.
2. a particle with a single negative charge; one or more electrons orbit the nucleus of the atom.

- **electron acceptor**

substance that accepts or receives electrons in an oxidation-reduction reaction, becoming reduced in the process.

- **electron carrier**

a molecule that conveys electrons; one of several membrane proteins in electron transport chains in cells. Electron carriers shuttle electrons during the redox reactions that release energy used to make ATP.

- **electron donor**
  substance that donates or gives up electrons in an oxidation-reduction reaction, becoming oxidised in the process.

- **electron microscope**
  a microscope that focusses an electron beam through a specimen, resulting in resolving power a thousand fold greater than that of a light microscope. A Transmission Electron Microscope (TEM) is used to study the internal structure of thin sections of cells. A Scanning Electron Microscope (SEM) is used to study the fine details of cell surfaces.

- **electron shell**
  an energy level at which an electron orbits the nucleus of an atom.

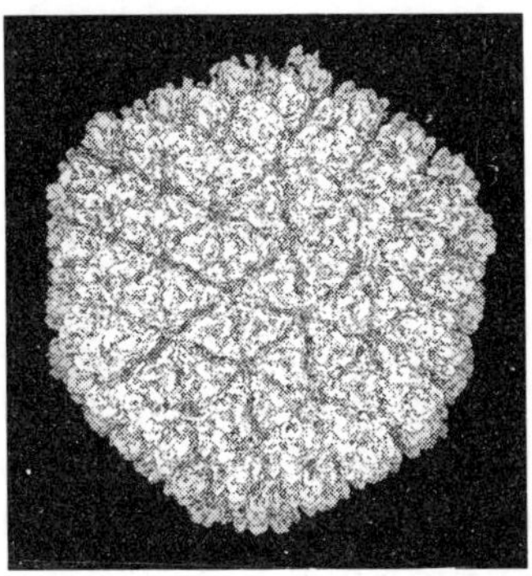

- **electron transport chain**
  1. a series of enzymes found in the inner membranes of mitochondria and chloroplasts. These are involved in transport of protons and electrons either across the membrane during ATP synthesis.
  2. a sequence of electron-carrier molecules (membrane proteins) that shuttle electrons during the redox reactions that release energy used to make ATP.

- **electronegativity**
  1. a property exhibited by some atoms whereby the nucleus has a tendency to pull electrons toward itself.
  2. the tendency for an atom to pull electrons towards it.

- **electronic charge unit**
  the charge of one electron (1.6021 x 10e-19 coulomb).

- **electrophoresis**
  a technique for separating molecules, particularly proteins, with different charges are separated by their different rates of migration in an electric field.

- **electrostatic force**
the attraction between particles with opposite charges.

- **electrostatic gradient**
the free-energy gradient created by a difference in charge between two points, generally the two sides of a membrane.

- **element**
1. any substance that cannot be broken down into another substance by ordinary chemical means.
2. any substance that cannot be broken down to any other substance.

- **enzyme**
a class of proteins serving as catalysts, chemical agents that change the rate of a reaction without being consumed by the reaction.

- **enzyme denaturation**
structural change in an enzyme, involving an unfolding of peptide chains and rendering the enzyme less soluble, produced by mild heat or various chemicals.

- **ephemeral**
lasting for a very short time.

- **ephemeral streams**
streams that flow only as the direct result of rainfall or snowmelt. They have no permanent flow.

- **ephemerals**
organisms with a short life cycle, especially plants whose seeds germinate, grow to produce new seeds and then die all in a short period within a year.

- **epibiont**
living attached to the surface of another organism but without benefit or deteriment to the host.

- **epicotyl**
a portion of the axis of a plant embryo above the point of attachment of the cotyledons; forms most of the shoot.

- **epideictic display**
a term for behavioural adaptations that have evolved to estimate population size of a species.

- **epidemic**
the outbreak of a disease which affects a large number and/or proportion of individuals in a population at the same time.

- **epidemiology**

1. the study of the occurrence of infectious diseases, their origins and pattern of spread through a population.
2. study of the patterns of disease or other harmful effects from toxic exposure within defined groups of people to find out why some people get sick and some do not.

- **epidermis**

1. the outermost portion of the skin or body wall of an animal.
2. the dermal tissue system in plants.
3. the outer covering of animals.

- **epididymis**

1. an elongated structure connected to the surface of the testis, it stores and matures sperm cells.
2. A long coiled tube into which sperm passes from the testis and are stored until matured and ejaculated.

Epididymis Tubule h.p.

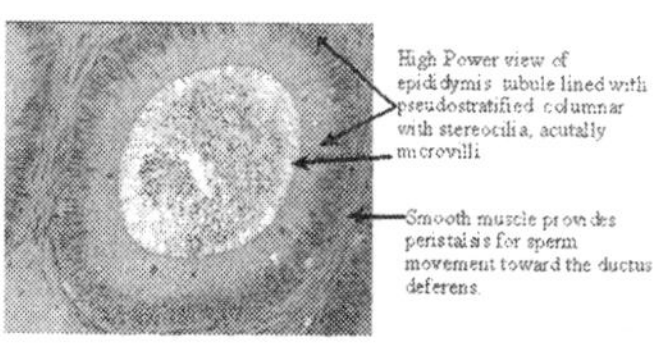

- **epigamic**

physical or behavioural features serving to attract or stimulate individuals of the opposite sex during courtship.

- **epigamic behaviour**

behaviour associated with mating including courtship displays. See **intersexual selection**.

- **epigenesis**

1. the developmental processes whereby the genotype is expressed as a phenotype. Biological development viewed as a process in which each stage in the growth and differentiation of an individual, from zygote to mature adult, arises from the preceding stage through the joint action of genetic and environmental determinants. In this view, at any point in development the phenotype already constructed sets the stage, as it were, for further construction, guided by the genes now brought into operation and environmental influences. Thus phenotype, genotype, and environment work in dynamic conjunc-

tion and progressive sequence. Epigenesis contrasts with performationism, according to which the organism is so to speak, prepackaged in miniature or in coded information in the zygote, which has only to be caused to grow for normal development to proceed.

Epigenesis Project: Halima Amer

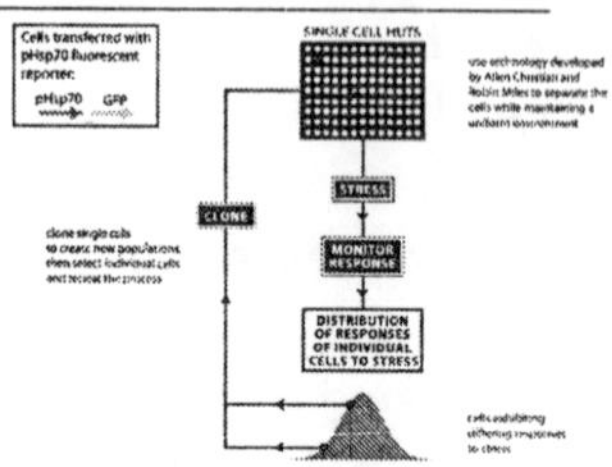

Classical ethology tended toward preformatio-nism; modern ethology is more inclined toward epigenesis.

2. the progressive development of form in an embryo.

3. a cartilaginous flap that blocks the top of the windpipe, the glottis, during swallowing, which prevents the entry of food or fluid into the respiratory system.

- **epigenetic**

developmental; pertaining especially to the interactions among developmental processes above the level of primary gene action.

- **epigenetic effects**

1. effects on phenotype that are due to gene-environment interactions.

2. explanations that go beyond a pure genetic determination of traits.

- **epinephrine**

1. a hormone produced as a response to stress; also called adrenaline. *See* **Adrenaline**.

- **epiphenomenon**

in English this word usually connotes an effect that is causally inconsequential and incidental or secondary to one that is operative or primary. For example, the squeaking sound that emanates from the turning of a pulley does not contribute anything to the work accomplished by the pulley. Mechanistic materialists have argued that mentality is an epiphenomenon and hence that we are deluded when we think, for example, that willing our actions causes them, since the only causes at work are physical ones, which produce mental states but are not affected by them. German ethologists use the word Epiphänomen to refer to an overt concomitant of a physiological state or process. For example, arousal or activation of the autonomic nervous system in situations of danger or threat can cause urination, defecation, trembling, feather ruffling, or hair erection. Such symptomatic expression has particular significance for ethology when it indicates the evolutionary origin of display behaviour. The signal function of the behaviour may be enhanced through ritualisation, and sometimes morphological support evolves. For example, many animals show vigorous tail wagging when aroused. In some species these movements have become threat displays, and in rattlesnakes and porcupines the tail waving also produced an acoustic signal, as a result of modification of structures (vertebrae in snakes, quills in porcupines). Also the excretion of urine for territorial marking is probably derived from an epiphenomenon-urination caused by 'anxiety'.

- **epiphyte**

1. a plant that nourishes itself but grows on the surface of another plant for support, usually on the branches or trunks of tropical trees.

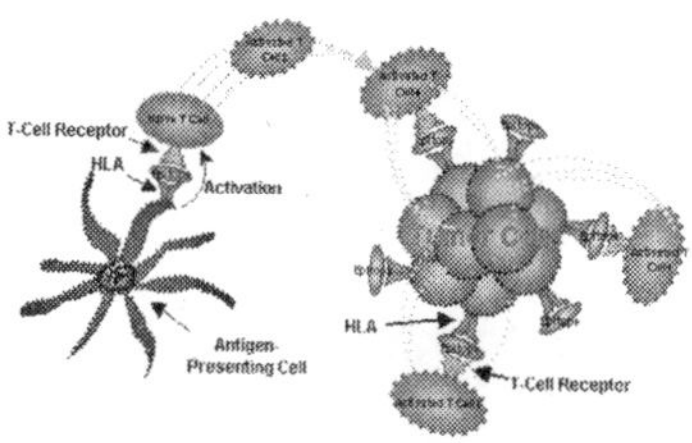

- **episome**

1. a circular gene fragment.
2. genetic element at times free in the cytoplasm, at other times integrated into a chromosome.
3. a plasmid capable of integrating into the bacterial chromosome.

- **epistasis**

1. a synergistic effect on the phenotype or fitness of two or more gene loci, whereby their joint effect differs from the sum of the loci taken separately.

2. a phenomenon in which one gene alters the expression of another gene that is independently inherited.
3. [Gk. a stopping]
4. a phenomenon in which one gene alters the expression of another gene that is independently inherited.

• **epistatic interactions**
reciprocal actions of genes at different loci that can modify the expression of one or the other or both sets of genes.

• **epistemology**
the study of the origins, methods, nature of and limits to knowledge.

• **epithelial tissue**
sheets of tightly packed cells that line organs and body cavities.

• **epithelium**
an animal tissue that forms the covering or lining of all free body surfaces, both external and internal.

• **epitope**
1. a localised region on the surface of an antigen that is chemically recognised by antibodies; also called antigenic determinant.
2. a localised region on the surface of an antigen that is chemically recognised by antibodies; also called antigenic determinant.

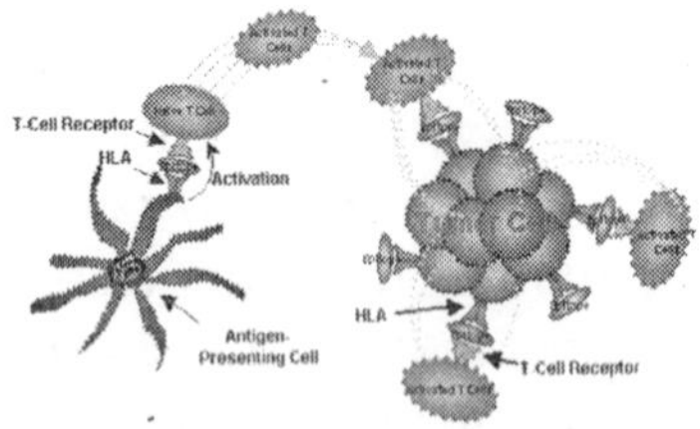

• **equation**
a precise representation of the outcome of a chemical reaction, showing the reactants and products, as well as the proportions of each.

• **equilibrium**
1. a condition of stasis, as of population size or genetic composition. When perturbed, the population tends to return to an equilibrium point.
2. in a reversible reaction, the point at which the rate of the forward reaction equals that of the reverse reaction. (constant) At equilibrium, the ratio of products to reactants. (potential) The membrane potential for a given ion at which the volt-

age exactly balances the chemical diffusion gradient for that ion.

3. the state of a system in which no further net change is occurring; result of counterbalancing forward and backward processes.

- **equilibrium species**
species characterised by low reproduction rates, long development times, large body size and long adult life with repeated reproductions.

- **equilibrium theory**
a theory of community organisation that focuses attention on the properties of the system at an equilibrium point, to which the community tends to return after a disturbance.

- **equitability**
the evenness with which individuals are distributed among species in a community.

- **equivalent variation**
the amount of money that leaves a person as well off as they would be after a change. Thus, it measures the amount of money required to maintain a person's satisfaction, or economic welfare, at the level it would be at after a change.

- **ergonomics**
the study of work, performance and efficiency or division of labour among animals.

- **erid**
emerging and re-emerging infectious diseases.

- **Ernst Mayer**
formulated the biological species concept.

- **erosion**
the wearing away of the land surface by wind or water.

- **erythrocyte**
1. a red blood corpuscle.
2. a red blood cell; contains hemoglobin, which functions in transporting oxygen in the circulatory system.

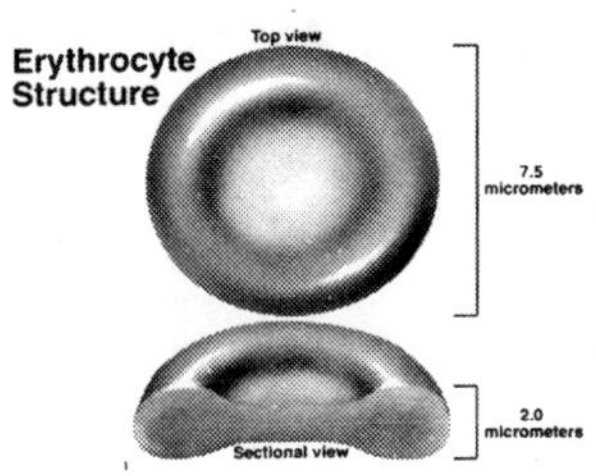

- **escape**
a form of defencive behaviour which may occur as soon as a predator is detected.

- **escape behaviour**
  rapid locomotion of an animal from any given location that occurs after specific stimuli are presented to it in that place.

- **escape cover**
  vegetation of sufficient size and density to hide an animal, or an area used by animals to escape from predators.

- **esophagus**
  1. an anterior part of the digestive tract; in mammals it leads from the pharynx to the stomach.
  2. a channel that conducts food, by peristalsis, from the pharynx to the stomach.

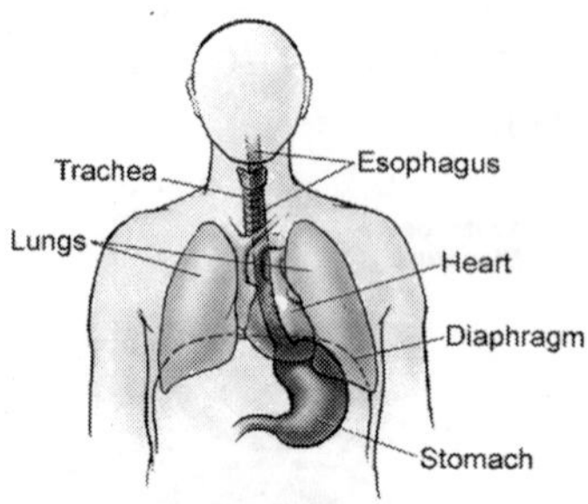

- **essential**
  1. an amino or fatty acid which is required in the diet of an animal because it cannot be synthesised.
  2. a chemical element required for a plant to grow from a seed and complete the life cycle.

- **essential amino acids**
  the amino acids that an animal cannot synthesize itself and must obtain from food. Eight amino acids are essential in the human adult.

- **estivation**
  a physiological state charac-terised by slow metabolism and inactivity, which permits survival during long periods of elevated temperature and diminished water supplies.

- **estivation**
  a physiological state characterised by slow metabolism and inactivity, which permits survival during long periods of elevated temperature and diminished water supplies.

- **estradiol**
  1,3,5(10)-estratriene-3,17 beta-diol $C_{18}H_{24}O_2$. This is the natural hormone - present in pure form in the urine of pregnant mares and in the ovaries of pigs.

• **estrogen**

1. a group of female hormones produced by the ovary.
2. any of a group of vertebrate female sex hormones.
3. The primary female steroid sex hormones, which are produced in the ovary by the developing follicle during the first half of the cycle and in smaller quantities by the corpus luteum during the second half. Estrogens stimulate the development and maintenance of the female reproductive system and secondary sex characteristics.

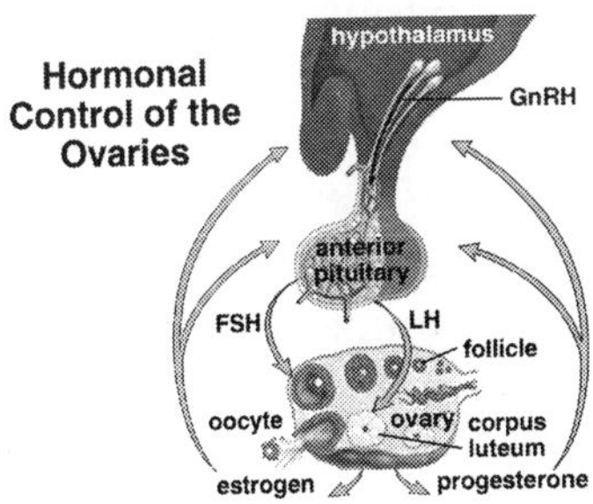

• **estrous cycle**

1. in female mammals, the higher primates excepted, a recurrent series of physiological and behavioural changes connected with reproduction.
2. a type of reproductive cycle in all female mammals except higher primates, in which the nonpregnant endometrium is reabsorbed rather than shed, and sexual response occurs only during midcycle at estrus.

• **estrus**

1. a regularly recurring period of ovulation and sexual activity by female mammals.
2. the limited period of heat or sexual receptivity that occurs around ovulation in female mammals having estrous cycles.

• **estuary**

that portion of a river that is close enough to the sea to be influenced by marine tides.

• **ethnobotany**

the study of indigenous knowledge bases regarding plants and their uses.

- **ethnopharmacology**
  the study of indigenous knowledge bases regarding medicines and how they are produced, as well as the medical practices, treatment protocols, etc. that utilise these medicines.

- **ethogram**
  1. an exact catalogue of all behavioural patterns occurring in a species including vocal patterns. An inventory of all action patterns.
  2. comprehensive compilation of the different behaviours exhibited by a species; A catalogue of actions as complete and precise as possible.

- **ethological isolation**
  reproductive isolating mechanisms that are due to behavioural preferences; potential mates meet but they do not choose to copulate with one another.

- **ethologist**
  1. an animal behaviourist who, typically, has been trained in zoology, usually studies the behaviour of insects, fishes and birds more often than that of mammals and other groups.
  2. student of comparative behaviour.

- **ethology**
  1. the study of animal behaviour from a zoological point of view. Ethology emphasizes species-typical behaviour, its adaptiveness, evolution, mechanisms and development. The ethologist applies the same methods to the study of behaviour that systematists, morphologists and embryologists use for the study of form. The study of the proximate mechanisms and adaptive value of animal behaviour.
  2. biological study of animal behaviour; comprises study of proximate causes of behaviour, including external stimulation and internal physiological mechanisms

and states; development, including the ontogenesis and genetics of behaviour; biological function, including immediate effects and ultimate consequences; and evolutionary origins and modifications.

3. The comparative study of patterns of animal behaviour, with emphasis on their adaptive significance and evolutionary origin.

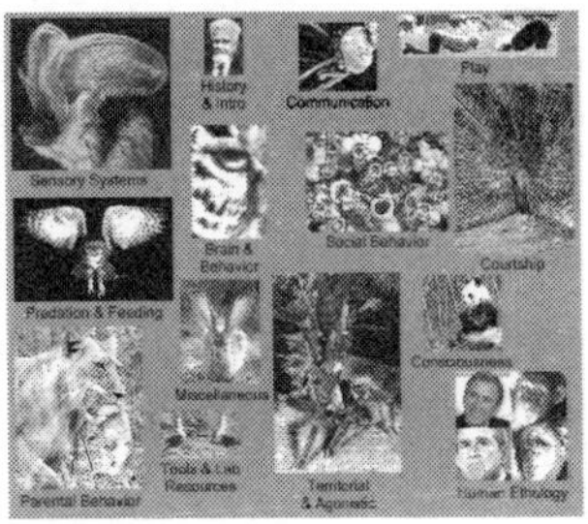

- **ethometry**

  term referring to the quantitative and functional analysis of releasing mechanisms and key stimuli.

- **ethomimicry**

  behavioural mimicry.

- **ethoparasite**

  various coleopterans have become social parasites of ants in that they live in the ant colony and are fed by the ants. a number of forms have succeeded by offering the appropriate key stimuli, in eliciting feeding behavior from ants.

- **ethospecies**

  species distinguished primarily by behavioural traits.

- **ethylene**

  1. the only gaseous plant hormone, responsible for fruit ripening, growth inhibition, leaf abscission and aging.

  2. the only gaseous plant hormone, responsible for fruit ripening, growth inhibition, leaf abscission and aging.

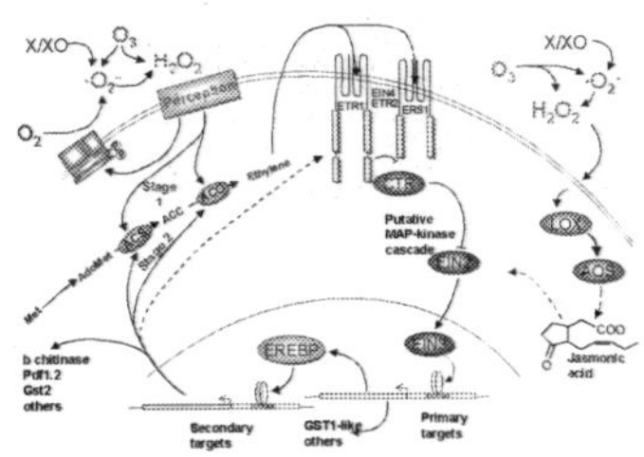

- **etiolation**

  In plants, a condition characterised by stem elongation, poor leaf development, and lack of chlorophyll; occurs in plants growing in the dark or with greatly reduced light.

- **eubacteria**

  the lineage of prokaryotes that includes the cyanobacteria and all other contemporary bacteria except archaebacteria.

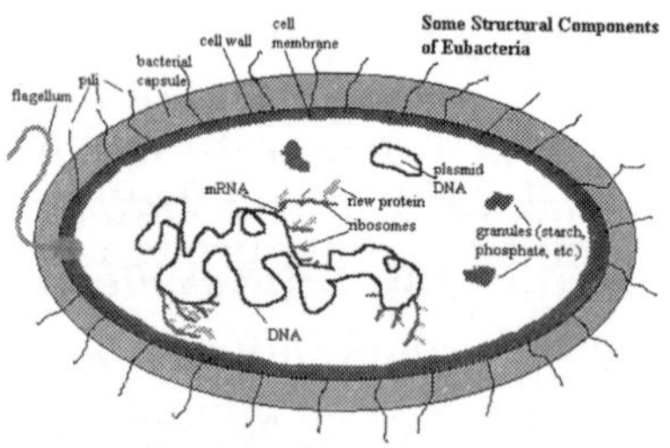

- **euchromatin**

  the more open, unravelled form of eukaryotic chromatin, which is available for transcription.

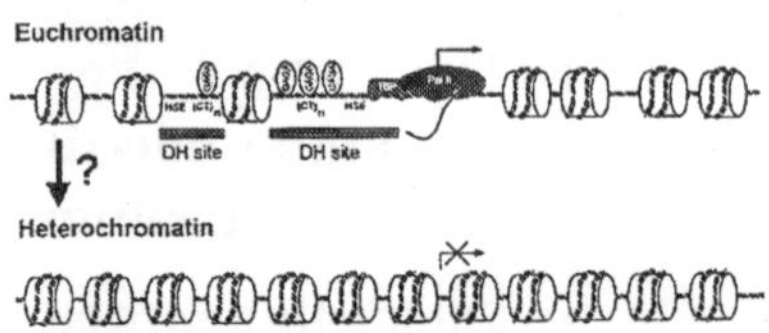

- **eucoelomate**

  an animal whose body cavity is completely lined by mesoderm, the layers of which connect dorsally and ventrally to form mesenteries.

- **eugenics**

  a program of decreasing the frequency of deleterious genes in a human population or of increasing that of advantageous genes through artificial selection against the genetically handicapped or in favour of the types considered especially desirable.

- **eukaryote**

  1. organism with cells possessing a membrane-bound nucleus in which the DNA is complexed with histones and organised into chromosome, i.e. protozoans, algae, fungi, plants and animals.
  2. cells having a well-defined nucleus and cellular organelles.
  3. a life form comprised of one or more cells containing a nucleus and membrane-bound organelles. Included are members of the Kingdoms Protista, Fungi, Plant and Animal.
  4. an organism whose cells contain membraine-bound organelles and whose DNA is enclosed in a cell nucleus and is associated with proteins.

- **eukaryotic cell**

  a type of cell with a membrane-enclosed nucleus and membrane-enclosed organelles, present in protists, plants, fungi and animals; also called eukaryote.

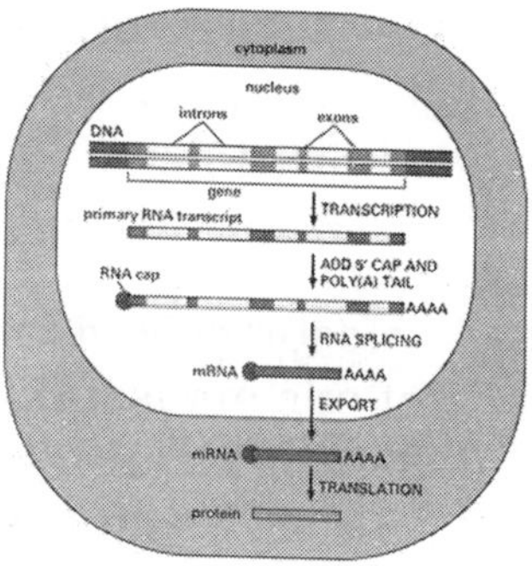

- **eumetazoa**
  members of the subkingdom that includes all animals except sponges.

- **eumetazoa**
  members of the subkingdom that includes all animals except sponges.

- **euphotic zone**
  the surface zone of a lake or ocean within which net primary productivity occurs.

- **eurytopic**
  of a species or population, capable of persistence in a wide variety of conditions or habitats; relative to a stenotopic species or population.

- **eusocial**
  1. means highly social and refers to the condition in which adult individuals of more than one generation are living together in a group, cooperating in nest construction or brood care and where there is some amount of reproductive dominance. In addition, highly eusocial species have permanent castes, i.e., where some individuals are sterile as adults.
  2. applied to animal societies, such as those of certain insects, in which sterile individuals work on behalf of reproductive individuals.

- **eutherian mammals**
  placental mammals; those whose young complete their embryonic development within the uterus, joined to the mother by the placenta.

- **eutrophic**
  a highly productive condition in aquatic environments which owes to excessive concentrations of nutrients which support the growth of primary producers.

• **eutrophic lake**
a highly productive lake, having a high rate of biological productivity supported by a high rate of nutrient cycling.

• **eutrophication**
1. enrichment of a water body with plant nutrients; usually resulting in a community dominated by phytoplankton.
2. a process in which an aquatic environment accumulates high nutrient levels due to factors such as industrial or urban pollution or run-off of fertilisers from nearby agricultural lands. The nutrients lead to dense blooms of algae and aquatic plants that cloud lake water, deplete specific minerals and dissolved gases and can cause natural plant and animal populations to decline.

• **evaginated**
folded or protruding outward.

• **evaporative cooling**
1. the property of a liquid whereby the surface becomes cooler during evaporation, owing to a loss of highly kinetic molecules to the gaseous state.

• **evapotranspiration**
the water loss to the atmosphere from soil and vegetation. The potential evapotranspiration may be calculated from physical features of the environmental such as incident radiation, wind speed and temperature. The evapotranspiration will commonly fall below the potential depending on the availability of water from precipitation and soil storage.

• **even aged management**
timber management actions that result in the creation of stands of trees in which the trees are essentially the same age.

• **even distribution**
the distribution of organisms in which the average distance between them is

greater than if they were distributed at random.

- **eversible**
  capable of being turned inside out.
- **evolution**
  1. processes by which living things on earth have diversified and changed. It involves changes in gene frequency within populations over time, which ultimately results in behavioural and phenotypic changes; related terms: Phylogeny.
  2. the process of change in the traits of organisms or populations over time. Evolution, through the process of natural selection, can lead to the formation of new species.
  3. All the changes that have transformed life on Earth from its earliest beginnings to the diversity that characterises it today.
  4. As used in biology, the term evolution means descent with change.
  5. (organic evolution) Any genetic change in a population of organisms from generation to generation; or more strictly, a change in allelic frequencies within populations from generation to generation. In a broad sense, the origin of entities possessing different states of one or more characteristics, and changes in their proportions over time. Organic evolution, or biological evolution, is a change over time of the proportions of individual organisms differing genetically in one or more traits; such changes transpire by the origin and subsequent alteration of the frequencies of alleles or genotypes from generation to generation within populations by the alterations of the proportions of genetically differentiated populations of a species.
- **evolutionarily stable strategy**
  a strategy which, if most members of a population adopt it, cannot be invaded by an alternative strategy. When two or more alternative ways of coping with a situation vie with one another, either in a population or in an individual's options, and each alternative is best

in some circumstances but not in all, the proportions in which the alternatives occur may reach an equilibrium determined by the combination that yields the highest net return in, for example, reproductive success. In a population in which reciprocal altruism has evolved, individuals that accept help without ever returning it ('spongers') will initially have an advantage over helpers because they will enjoy the benefits of the altruistic system without incurring any of the costs. Spongers will therefore be selected: they will increase in the population at the expense of the helpers. But as the proportion of helpers decreases, sponging becomes less and less profitable. At some point the advantage the remaining helpers get from one another outweighs the advantage of being exploited, and they make a comeback. This will be checked as their recovery provides an advantage to the spongers. The proportions of helpers and spongers will eventually converge to an equilibrium ratio at which net reproductive success is maximal; that ratio of spongers to helpers becomes the ESS. Any shift away from it results in decreasing success for the increasing variant. This argument, derived from mathematical games theory, has been applied to such matters as the evolution of threat display and parental investment.

- **Evolutionarily Stable Strategy (ESS)**
  a strategy which, if adopted by most of a population, cannot be bettered by any other strategy, and will therefore tend to become established by natural selection.

- **evolutionary clocks**
  point mutations occur and are fixed in the genome at a reasonably regular rate. An evolutionary clock uses these substitutions as a measure of the amount of time that has passed between the divergence of two lineages from each other.

- **evolutionary conflict of interest**
  the situation in which those traits that maximise the reproductive success of one individual cause a reduction

in the reproductive success of another. Evolutionary conflicts of interest are common between species such as predators and prey or parasites and hosts, but they also occur within a species, such as between competitors for food or females. Evolutionary conflicts of interest may even occur between closely related individuals such as parents and offspring or the members of a pairs.

- **evolutionary species concept**
  the idea that evolutionary lineages and ecological roles can form the basis of species identification.

- **evolutionary trees**
  lineages designed to show the evolutionary history of relationships between groups of organisms.

- **evolutionary stable strategy**
  strategy that when common cannot be invaded by an alternative (mutant) strategy under the influence of natural selection. i.e. a strategy that performs better than any other strategy against itself.

- **exact compensation**
  density dependence in which increases in initial density are exactly counterbalanced by increases in death rate and/or decreases in birth rate and/or growth rate, such that the outcome is the same irrespective of initial density.

- **exafference**
  afference resulting from stimulation of exteroceptors through environmental change with respect to which the animal is passive; for example, a cloud passing in front of the sun will cause the animal to perceive a drop in light intensity not brought about by its own activity.

- **exaptation**
  1. same as preadaptation except it does not imply foresight.
  2. a structure that evolves and functions in one environmental context but that can perform additional functions when placed in some new environment.

- **excitability**
  excitability is a basic property of living organisms in which changes in state elicits an arousal response. Re-

fers to an endogenous production of excitation of the CNS occurring even in deafferented animals.

- **excitable cells**
  a cell, such as a neuron or a muscle cell, that can use changes in its membrane potential to conduct signals.

- **excitation**
  a hypothetical state of the animal used in various ways to account for the occurrence of a response and presumed to be great when response magnitude is large.

- **excitatory postsynaptic potential**
  1. potential that is generated in a postsynaptic neuron by activation of an excitatory synapse, usually depolarising.
  2. an electrical change (depolarisation) in the membrane of a postsynaptic neuron caused by the binding of an excitatory neurotransmitter from a presynaptic cell to a postsynaptic receptor. This phenomenon facilitates generation of an action potential in the PSP.

- **Excitatory Postsynaptic Potential (EPSP)**
  an electrical change (depolarisation) in the membrane of a postsynaptic neuron caused by the binding of an excitatory neurotransmitter from a presynaptic cell to a postsynaptic receptor; makes it more likely for a postsynaptic neuron to generate an action potential.

- **excitatory transmission**
  synaptic transmission that has a depolarising effect on the postsynaptic neuron and which increases the probability that the postsynaptic neuron will fire an action potential or the amount of neurotransmitter release.

- **exclusion principle**
  the principle stating that two species cannot coexist at the same locality if they have identical ecological requirements.

- **excretion**
  1. the disposal of nitrogen-containing waste products of metabolism.
  2. release of materials which arise in the body due to metabolism (e.g., CO_2, NH_3, H_2O).

- **excretory system**
  the organ system that disposes of nitrogen-containing metabolic wastes.

- **exergonic**
  a phenomenon which involves the release of energy.

- **exergonic reaction**
  spontaneous chemical reaction in which there is a net release of free energy.

- **exhaustion**
  of action-specific energy, or of specific action potential, a hypothetical process postulated to account for the data of adaptation or habituation of a species-specific response.

- **existence value**
  the value that people place on simply knowing that something exists, even if they will never see it or use it.

- **exocrine glands**
  glands, such as sweat glands and digestive glands, that secrete their products into ducts that empty onto surfaces, such as the skin, or into cavities, such as the interior of the stomach.

- **exocytosis**
  1. The cellular secretion of macromolecules by the fusion of vesicles with the plasma membrane.
  2. a process by which a vesicle within a cell fuses with the plasma membrane and releases its contents to the outside.

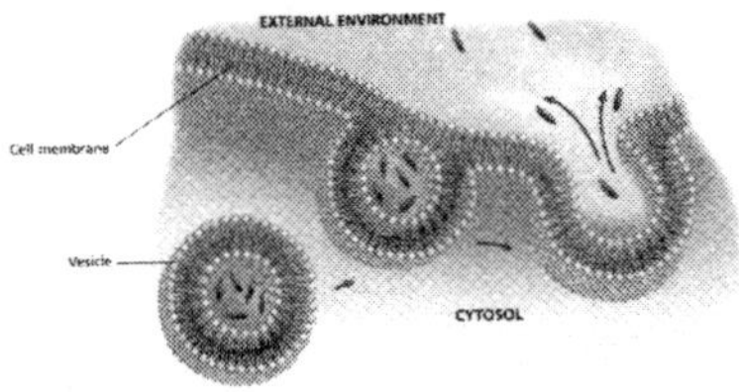

- **exogenous**
  1. originating outside of an organism.
  2. from outside the organism.

- **exon**
  1. a part of an interrupted gene that is translated into a polypeptide.
  2. the coding region of a eukaryotic gene that is expressed. Exons are separated from each other by introns.
  3. a part of a primary transcript (and the corresponding part of a gene) that is ultimately either translated (in the case of mRNA) or uti-

lized in a final product, such as tRNA.

- **exoskeleton**

1. a hard encasement on the surface of an animal, such as the shells of mollusks or the cuticles of arthropods, that provides protection and points of attachment for muscles.
2. an external skeleton, characteristic of members of the phylum, Arthropoda.

- **exotoxin**

1. A toxic protein secreted by a bacterial cell that produces specific symptoms even in the absence of the bacterium.

- **experience**

experience arises through storage of learned contingencies.

- **exploded lek**

sites where individual males display and court females, usually associated with a specific environmental feature such as a drumming log or display perch.

- **exploitation competition**

competition in which any adverse effects on an organism are brought about by reductions in resource levels caused by other competing organisms.

- **exploiter-mediated coexistence**

a situation where predation promotes the coexistence of species amongst which there would otherwise be competitive exclusion.

- **exploratory behaviour**

behaviour which produces the species-typical orientation in time and space, necessary for effective learning.

- **explosive breeding assemblage**

the formation of large groups of reproducing individuals in species that breed on only a few days each year.

- **exponential**

(population growth) the geometric increase of a population as it grows in an ideal, unlimited environment.

- **exponential growth**

1. growth in the size of a population (or other entity) in which the rate of growth increases as the size of the population increases.
2. a phase of growth of a population in which the time of doubling in size is con-

stant. Under exponential growth the log of the number of individuals increases linearly with time.

3. in populations, the increasingly accelerated rate of growth due to the increasing number of individuals being added to the reproductive base. Exponential growth is very seldom approached or sustained in natural populations.

4. growth in which some quantity, such as population size or economic output, increases by a fixed percentage of the whole in a given time; when the increase in quantity over a long enough time is plotted, this type of growth typically yields a curve shaped like the letter J.

- **expression vector**

a vector that allows a DNA sequence cloned into it to be transcribed when the vector is introduced into a cell.

- **expressivity**

in genetics, the degree to which a particular genotype is expressed in the phenotype of individuals with that genotype.

- **extant**

a species that is still alive and reproducing. All species that currently live on earth are extant.

- **extended family**

a family containing kin beyond the nuclear family.

- **external fertilisation**

a form of sperm transfer in which eggs and sperm are united outside the body.

- **externalities**

uncompensated side effects of human actions. For example, if a stream is polluted by runoff from agricultural land, the people downstream experience a negative externality.

- **exteroreceptors**

sense organs which perceive stimuli emanating from outside the body: eyes, ears, sense of touch, warmth and cold receptors, current and electrical receptors; in contrast to receptors that respond to internal information such as proprioreceptors.

- **extinct**

1. refers to a species that no longer exists. Local extinction occurs when every

member of a particular population has died. Global extinction occurs when every member of a species has died. The passenger pigeon and the dodo are examples of globally extinct birds.
2. no longer existing.
3. a species that is no longer living on earth. All representatives of the species are dead. All the species that once occupied the earth but are no longer living are extinct. We know of their existence through studying the *fossil record*. Compare to **extant**.

- **extinction**

1. the condition that arises from the death of the last surviving individual of a species, group, or gene, globally or locally.
2. termination of a lineage. The process by which learned behaviour patterns cease to be performed when they are no longer appropriate.
3. complete disappearance of a species from the earth. This happens when a species cannot adapt and successfully reproduce under new environmental conditions, when it evolves (through a process called *radiation*) into one or more new species, or when every member of the species is killed by overpredation, pollution, or other man-made causes.

- **extinguish**

1. to omit reinforcement of a response sufficiently often that a decrement in response-strength is observed.
2. with reference to a response, to decrease in magnitude with the omission of reinforcement.

- **Extra Cellular Matrix (ECM)**

the substance in which animal tissue cells are embedded; consists of protein and polysaccharides.

- **extraembryonic membranes**

four membranes (yolk sac, amnion, chorion, allantois) that support the developing embryo in reptiles, birds and mammals.

- **extrafloral nectaries**

nectar secreting glands found on the leaves and other vegetative parts of plants.

- **extra-pair copulation**

refers to copulations that occur between individuals

that do not have a pair bond (in species in which pair bonding is the norm.) Now often abbreviated EPC.

- **extravagant character**
these are characters that have evolved by female choice and the runaway process. They are usually characters which have an obvious natural selective disadvantage but are maintained due to sexual selection.

- **extrinsic**
external to, not a basic part of; as in extrinsic isolating mechanism.

- **extrinsic factors**
literally, factors acting from outside. In ecology, physical and chemical features of the environment, and other organisms, are all extrinsic factors acting on an organism.

- **eye contact**
a behavioral element in the social interactions of some animals.

- **eye spots**
in many Lepidopterans and some Homopterans eye spots are present on the rear wings. These spots are normally covered when the animal is resting, but are displayed when the animal is disturbed as it quickly raises its forewings. The eye spots strongly resemble mammalian eyes.

- **eye-brow flash**
in human greeting gestures, the raising of the eye-brows followed often by a smile.

- **eye-camouflage**
the eyes of many species of fishes and other animals are obscured by a black eye bar.

- **f plasmid**
the fertility factor in bacteria, a plasmid that confers the ability to form pili for conjugation and associated functions required for transfer of DNA from donor to recipient.

- **f_1 generation**
first filial generation. The progeny resulting from a parental cross between two pure lines.

- **f_2 generation**
1. second filial generation, resulting from the mating of two F1 individuals.
2. offspring resulting from interbreeding of the hybrid F1 generation.

- **facies**
the representation of a biome in the sediments, e.g.,

marine facies are sediments of marine origin.

- **facilitated diffusion**
  passive movement through a membrane involving a specific carrier protein; does not proceed against a concentration gradient.

- **facilitation**
  1. increase in the amplitude of postsynaptic responses as a result of repetitive activation of the synapse.
  2. in which the influence of early species in a community succession is to facilitate establishment of later ones by changing the conditions encountered.

- **facing away**
  appeasement where head is averted from opponent.

- **factor analysis**
  a statistical method for analysing the degrees of covariation among a number of different measures of a capacity such as intelligence or of a type of behaviour such as courtship. Factor analysis involves working out the partial correlation coefficients between the different combinations of variables and treating them in such a way as to reveal dependence on a small number of common factors. Ethologists have used this and similar techniques (hierarchical cluster analysis, multivariate analysis, principal component analysis) to generate evidence for the existence and organisation of drives or motivational systems underlying behavioural patterning.

- **facultative**
  1. an organism which exhibits the capability of changing from one habit or metabolic pathway to another, when conditions warrant.
  2. An organism that makes ATP by aerobic respiration if oxygen is present but that switches to fermentation under anaerobic conditions.

- **facultative annual**
  an organism which may in some circumstances complete its life cycle within 12 months. The phrase is usually used of plants.

- **facultative mutualism**
  the condition in which one or both species in a mutu-

alistic association may survive and maintain populations in the absence of the other partner.

- **fertilisation**
  fusion of nuclei of egg and sperm.

- **foetus**
  an embryo in its later stages of development, still in the egg or uterus.

- **fibre**
  a lignified cell type that reinforces the xylem of angiosperms and functions in mechanical support; a slender, tapered sclerenchyma cell that usually occurs in bundles.

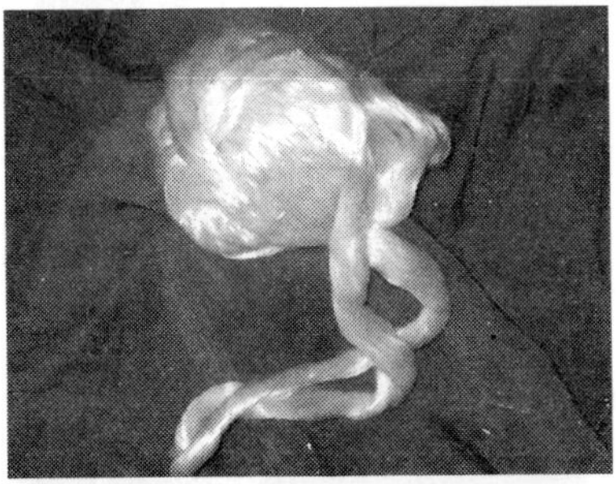

- **fibrin**
  the activated form of the blood-clotting protein fibrinogen, which aggregates into threads that form the fabric of the clot.

- **fibroblast**
  a type of cell in loose connective tissue that secretes the protein ingredients of the extracellular fibres.

- **fibronectins**
  a family of extracellular glycoproteins that helps embryonic cells adhere to their substrate as they migrate.

- **field capacity**
  the condition of a soil when it contains all the water retained after draining freely under gravity.

- **fighting**
  a physical conflict between two individuals.

- **filial cannibalism**
  cannibalism of kin particularly siblings.

- **filial imprinting**
  see **Following response.**

- **filter**
  high-order neurons in a sensory system which transmit only certain aspects of the signals they receive.

- **final common path**
  competition for effect or output when the animal is faced with motivational conflict.

- **final cut**
  the removal of the last seed bearers or shelter trees after regeneration of new trees has been established in a stand being managed under the shelter wood system of silviculture.

- **fine filter management**
  management that focuses on the welfare of a single or only a few species rather than the broader habitat or ecosystem.

- **fine-grained environment**
  the environment experienced by an organism with good homeostatic mechanisms, high behavioural plasticity and mobility; experienced as a succession of interconnected conditions.

- **fire cycle**
  the average time between fires in a given area.

- **fire regime**
  the characteristics of fire in a given ecosystem, such as the frequency, predictability, intensity, and seasonality of fire.

- **first law of human ecology**
  we can never do merely one thing. Any intrusion into nature has numerous effects, many of which are unpredictable. For example, one classic dilemma is the case of behavioural biologists who observe their study subjects at close range: Are the observed behaviours truly natural or are they influenced by the researcher's presence?

- **fisheries habitat**
  streams, lakes, and reservoirs that support fish or have the potential to support fish.

- **fission**
  1. one chromosome breaks into two.
  2. social groups that divide.

- **fitness**
  1. the contribution made to a population of descendants

by an individual relative to the contribution made by others in its present population. The relative contribution that an individual makes to the gene pool of the next generation.
2. probability that an animal of a particular genotype and phenotype will reproduce; in game theory models used to judge the success of a strategy (commonly denoted as (w) or $E_I(J)$ = expected pay-off to I when playing against J); see also **Inclusive Fitness, Direct Fitness, Indirect Fitness.**
3. the extent to which an individual passes on its genes to the next generation. Relative fitness is the number of offspring of an individual compared to the mean.
4. the average contribution of one allele or genotype to the next generation or to succeeding generations, compared with that of other alleles or genotypes.
5. in population genetics, W, the rate of increase of one genotype (Aa, aa) relative to another (AA); S=1-W where S is the selection coefficient. See **inclusive fitness, individual fitness, population fitness; direct fitness, indirect fitness.**

- **fitness benefit**
  that aspect of a trait that tends to raise the inclusive fitness of individuals.
- **fitness cost**
  that aspect of a trait that tends to decrease inclusive fitness of individuals.
- **fixation**
  1. attainment of a frequency of 1 (i.e. 100%) by an allele in a population, which thereby becomes monomorphic for the allele.
  2. conversion of a substance into a biologically more usable form, for example, CO_2 fixation during photosynthesis and N_2 fixation.
  3. process of treating living tissue for microscopic examination.

- **Fixed Action Pattern (FAP)**

1. translated from German meaning 'inherited movement coordination' In the behaviour of an animal one repeatedly observes certain movements which have a constant form and which need not be learned and those movements are characteristic of the species.
2. FAP refers to a kind of genetically fixed inborn ability.

- **fixed costs**

production costs that are not related to the level of production; also referred to as overhead costs.

- **flaccid**

limp; walled cells are flaccid in isotonic surroundings, where there is no tendency for water to enter.

- **flagellin**

the protein from which prokaryotic flagella are constructed.

- **flagellum**

a long whip like appendage that propels cells during locomotion in liquid solutions. The prokaryote flagellum is comprised of a protein, flagellin. The eukaryote flagellum is longer than a cilium, but as a similar internal structure of microtubules.

- **flame cell**

a flagellated cell associated with the simplest tubular excretory system, present in flatworms- it acts to directly regulate the contents of the extracellular fluid.

- **fledging**

when a young bird leaves its parents nest.

- **flehmen**

a widely distributed behaviour pattern among mammals in which the mouth is opened, the teeth exposed by retracting the upper lip (lip curl), the nasal openings closed and the head jerked back. It occurs particularly in the reproductive season after the male has licked the females urine or vulva. (H) This behaviour is associated with the operation of the vomeronasal organ.

- **flexibility**

the ability to change over the life of an individual.

- **flight behaviour**
  1. flying of birds and bats.
  2. rapid locomotion of an animal from proximity to another animal that is exhibiting threat or attack behaviour.
- **flight distance**
  the distance at which an animal will flee from a predator or rival. Flight distance varies from species to species and individual to individual.
- **floater**
  a non-breeder that is a potential breeder and is not bound to a particular territory.
- **flock**
  a number of animals travelling in company, commonly of the same or similar species.
- **flocking**
  a group of birds that remains together as a result of social attraction between individuals.
- **flood plain**
  a lowland adjoining a water-course. At a minimum, the area is subject to a 1% or greater chance of flooding in a given year.
- **flora**
  1. the plant life of an area.
  2. the plants that live in a particular area.
  3. the plants of a given area or period.

- **floristic equilibrium**
  the species composition of a flora when there is no further internally or externally generated change.
- **flow diagram**
  a block and arrow diagram showing the sequencing of transitions from one pattern to the next.
- **flow field**
  pattern of coherent motion associated with a particular kind of movement.
- **fluid feeder**
  an animal that lives by sucking nutrient-rich fluids from another living organism.
- **fluid mosaic model**
  the currently accepted model of cell membrane structure, which envisions

the membrane as a mosaic of individually inserted protein molecules drifting laterally in a fluid bilayers of phospholipids.

- **focal animal sampling**
  a method for taking observational data in which one individual is observed for long periods of time before proceeding to the next animal.

- **folivores**
  animals that eat leaves.

- **follicle**
  a jacket of cells around an egg cell in an ovary.

- **Follicle Stimulating Hormone (FSH)**
  a gonadotropic hormone of the anterior pituitary that stimulates growth of follicles in the ovaries of females and function of the seminiferous tubules in males.

- **following response**
  the innate disposition of young nidifugous birds to approach and follow a moving object shortly after hatching. This moving companion may be but does not have to be acquired during a short sensitive period in early life. The early experience may affect later preferences. For example, a jackdaw hand-raised by Lorenz considered humans as parent or sexual companions throughout its life.

- **food chain**
  1. an abstract representation of the links between consumers and consumed populations, e.g. plant - herbivore - carnivore.
  2. sequence of organisms, including producers, consumers, and decomposers, through which energy and materials may move in a community.
  3. figure of speech describing the dependence of heterotrophs on other organisms for food, progressing in a series beginning with primary producers (plants) and ending with the largest carnivores. The food chain is used as a figurative image for educational purposes only... real trophic systems resemble webs rather than chains. See **food web**.

- **food web**

1. representation of feeding relationships in a community that includes all the links revealed by dietary analysis.
2. the elaborate, interconnected feeding relationships in an ecosystem.
3. the combination of all the feeding relationships that exist in an ecosystem. Most prey species are eaten by many different predators, and most predators eat more than one prey item. As a result, a picture of a trophic system with lines (representing ecological relationships) drawn between predators and prey soon resembles an intricate web.

- **food-begging**

begging by the young for food from the parents. In courtship a ritualised form of begging may occur between partners. see **courtship feeding**.

- **food-running**

distinctive rapid locomotion most frequently displayed by young birds in the presence of peers after grasping worms or worm-shaped objects with their beaks. A bird running with such an object in its beak is followed by peers who attempt to steal the object.

- **foot candle**

unit of illumination; the illumination of a surface produced by one standard candle at a distance of one foot.

- **forage**

all browse and non-woody plants that are eaten by wildlife and livestock.

- **foraging**

refers to behaviour associated with searching for subduing, capturing and consuming food.

- **foraging strategy**

the methods used by an animal in its search for food.

- **forb**

1. a broadleaf plant that has little or no woody material in it.
2. a broad-leafed (dicot) herbaceous plant. The term is especially applied to such plants living in grasslands.

- **forced copulation**
rape, in which a male inseminates a female against her will and with the consequence that her fitness is reduced on average.

- **foreground**
the part of a scene or landscape that is nearest to the viewer.

- **foreleg kick**
several species of male ruminants will use a stiff-legged, forward-directed kick with their forelimb as a part of the courtship behaviour. The forelimb may be raised between the female's hind limbs or to the side of them.

- **forest health**
a measure of the robustness of forest ecosystems. Aspects of forest health include biological diversity; soil, air, and water productivity; natural disturbances; and the capacity of the forest to provide a sustaining flow of goods and services for people.

- **forest roads and trails**
roads and trails under the jurisdiction of the Forest Service.

- **forest supervisor**
the official responsible for administering National Forest lands on an administrative unit, usually one or more National Forests. The forest supervisor reports to the regional forester.

- **formula**
a precise representation of the structure of a molecule or ion, showing the proportion of atoms which comprise the material.

- **fossil**
1. any evidence of the former existence of a living thing, including bones, casts, molds, etc.
2. a relic or impression of an organism from the past, preserved in rock.
3. a remnant, impression, mineralised mold, amber encasement, or other trace of a once-living organism. Technically, anything that once lived and has been permanently preserved is a fossil, but the most common usage implies great age. This common usage of fossil generally refers to the mineralised remains or impressions, preserved in stone (almost al-

ways sedimentary rock), of extinct organisms from past geologic ages.

- **fossil fuel**
  products of partial of complete decomposition of plants and animals that occur as crude oil, coal, natural gas, or heavy oils as a result of exposure to heat and pressure in earth's crust over millions of years.

- **fossil record**
  the cumulative taxonomic information and historical perspective provided by the wealth and diversity of fossils and related geologic data stored in the earth's crust.

- **foster rearing**
  the raising of young by individuals other than the natural or biological parents. If this is done by individuals of a different species, it may be refereed to as interspecific fostering, or cross-fostering, or cross-species rearing. When young animals are raised by people it is called hand-raising. As a rule only human contrived situations are referred to as foster rearing; hence brood parasitism is usually excluded and if parental substitution occurs in other natural conditions, the more common term is 'adoption'.

- **founder effect**
  1. the principle that the founders carry a fraction of the total genetic variation of the source population; when a new population is established in isolation, its gene pool is not identical to that of the parent population because of sampling error. Changes in gene frequencies in a population through genetic drift which is caused by a few animals founding a new population.
  2. the difference between the gene pool of a population as a whole and that of a newly isolated population of the same species.

- **fovea**
  in the vertebrate retina, the region specialised for high visual acuity.

- **fractionation**
  an experimental technique which involves separation of parts of living tissue from one another using centrifugation.

- **fragmentation**

1. the splitting or isolating of patches of similar habitat, typically forest cover, but including other types of habitat. Habitat can be fragmented naturally or from forest management activities, such as clear cut logging.
2. the breaking up of large habitats into smaller, isolated chunks. Fragmentation is one of the main forms of habitat destruction, which is the primary reason biodiversity is in decline.
3. a mechanism of asexual reproduction in which the parent plant or animal separates into parts that reform whole organisms.

- **frameshift mutation**

a mutation resulting from the insertion or deletion in DNA of one or more bases, thus changing the boundaries of the codons and altering the whole string of amino acids subsequent to the insertion or deletion.

- **fraser darling effect**

acceleration and synchronization of reproductive state in a breeding colony as a consequence of communal social stimulation. The effect is greater as group size or density increases.

- **free energy**

usable energy in a chemical system; energy available for producing change.

- **freemartin**

the female twin of a bull calf. Freemartins are usually infertile.

- **free-running cycle**

the cycle of activity of an individual that is expressed in a constant environment, i.e. without entertainment cues.

- **freezing behaviour**

adoption of a fixed, immobile stance usually in an anti-predator context.

- **frequency**

reciprocal of the wavelength (the interval from a given point on one sound wave to the equivalent point on the next sound wave), unit: Hertz = cycles/s.

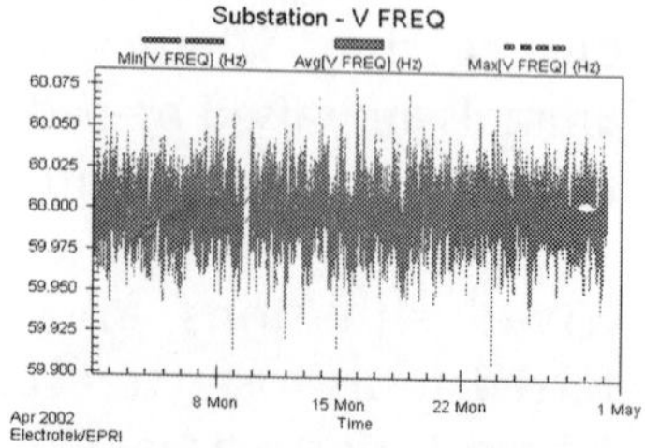

- **frequency code**
  information transfer based on the rate of action potentials reaching more than 100/s for intense stimuli.
- **frequency dependence**
  of a predator, describes the tendency to take disproportionately more of the commoner prey species.
- **frequency distribution**
  a synthetic presentation of a series of observations, obtained by specifying the number of observations falling in each class.
- **frequency modulation**
  variation of the wavelength of energy transmitted in wave form. In bioacoustics and animal behaviour literature the term usually refers to sound, the frequency modulation of which is experienced as change of pitch. Melody is thus an effect of frequency modulation.
- **frequency-dependent selection**
  1. selection in which the fitness of a phenotype is a function of its frequency in the population.
  2. a decline in the reproductive success of a morph resulting from the morph's phenotype becoming too common in a population; a cause of balanced polymorphism in populations.
- **fright**
  the feeling of being threatened with its accompanying alertness and acumen. Fright unlike fear is object-oriented, i.e. it occurs only in response to tangible danger
- **fright response**
  the escape response of animals when in physical danger.
- **frost heave**
  a land surface that is pushed up by the accumulation of ice in the underlying soil.
- **fruit**
  1. a mature ovary or cluster of ovaries (sometimes with additional structures associated with the ovary).
  2. (aggregate) a ripened ovary, or fruit, resulting from a single flower with several separate carpels.
  3. (simple) a fruit derived from a single ovary.

- **fruiting body**
  a spore-bearing structure (e.g., the aboveground portion of a mushroom.)

- **frustration**
  behaviour, often appearing to be inappropriate, resulting from a failure of an expected event to occur which itself would satisfy some primary or secondary need. The operation of preventing an animal from making some response. This may be done in any of three ways:
  1. by withholding the stimulus for the response, when the stimulus ordinarily appears as the consequence of a previous response; or.
  2. by mechanically preventing the response from occurring.
  3. a state of motivation that arises in situations in which the consequences of behaviour are less than those which the animal has been led to expect on the basis of past experience.

- **fuels**
  plants and woody vegetation, both living and dead, that are capable of burning.

- **fuels management**
  the treatment of fuels that would otherwise interfere with effective fire management or control. For instance, prescribed fire can reduce the amount of fuels that accumulate on the forest floor before the fuels become so heavy that a natural wildfire in the area would be explosive and impossible to control.

- **fuel wood**
  wood cut into short lengths for burning.

- **fugitive species**
  one that occupies temporary environments or habitats and so does not persist for many generations at any one site.

- **function**
  1. as used in everyday language function refers to the job that something is designed to do. Biologist use function in the more specialized sense of the adaptive significance of the character or how that character increases reproductive success of the individual.
  2. all the processes within an ecosystem through

which the elements interact, such as succession, the food chain, fire, weather and the hydrologic cycle.

- **functional group**
one of several groups of atoms commonly found in organic molecules. A functional group contributes somewhat predictable properties to the molecules which possess them.

- **functional response**
the relationship between a predator's consumption rate of prey and the density of those prey.

- **fundamental net reproductive rate**
the average number of individuals that each existing individual in a population gives rise to one time interval later; hence, the multiplication rate relating size of a population to its size one time interval earlier.

- **fundamental niche**
1. the largest ecological niche that an organism or species can occupy in the absence of interspecific competition and predation.
2. set of resources and physical factors required for survival and reproduction of individuals of a species.

- **fundamental theorem of natural selection**
formulated by R.A. Fisher and states that the rate of evolution at the level of the individual increases with genotypic variance.

- **fusion**
1. two chromosomes fuse into one.
2. two social groups join.

- **g**
(protein) a membrane protein that serves as an intermediary between hormone receptors and the enzyme adenylate cyclase, which converts ATP to cAMP in the second messenger system in nonsteroid hormone action. Depending on the system, G proteins either increase or decrease cAMP production.

- **g1 phase**
the first growth phase of the cell cycle, consisting of the portion of interphase before DNA synthesis is initiated.

- **g2 phase**
the second growth phase of the cell cycle, consisting of the portion of interphase

after DNA synthesis but before mitosis.

- **gaia hypothesis**
  1. the theory that the earth functions as a single independent living organism.
  2. an idea, which suggests that the biosphere of the earth exists as a 'superorganism' which exhibits homeostatic self-regulation of the environment-biota global system.

- **gallop**
  the fastest naturally developed gait. The leg movement is similar to the canter except the diagonal beat is extended so the fore hoof hits the ground slightly later than the hind hoof and the phase of total suspension in the air is longer.

- **gamboling**
  bouncing and turning, seemingly stiff-legged; seen in young sheep and goats when playing.

- **game species**
  any species of wildlife or fish that is harvested according to prescribed limits and seasons.

- **game theory**
  a part of applied mathematics dealing with the situation in which the probabilities of winning and losing vary with the strategies chosen by the players. See **ESS**.

- **gametangium**
  the reproductive organ of bryophytes, consisting of the male antheridium and female archegonium; a multichambered jacket of sterile cells in which gametes are formed.

- **gamete**
  1. a reproductive cell (haploid) which unites with another in fertilization to produce a zygote from which a new individual (genet) arises.
  2. a mature male or female reproductive cell (sperm or egg).
  3. a sexual reproductive cell that must usually fuse with another such cell before development begins; an egg or sperm.

- **gametic selection**
  differential survival of sperm or egg cells.

- **gametogenesis**

the combined process of cell division and differentiation which results in the production of gametes.

- **gametophyte**

a haploid plant that can produce gametes.

- **ganglion**

1. a dense cluster of many neurons in the central nervous system.
2. discrete mass of cell bodies and neurites surrounded by a connective tissue sheath. In invertebrates, the central nervous system frequently consists of a chain of ganglia joined by connectives. In vertebrates, ganglia are always located outside the central nervous system. Also an obsolete term for a brain nucleus, now used only for basal ganglia.
3. a structure containing a group of cell bodies of neurons.

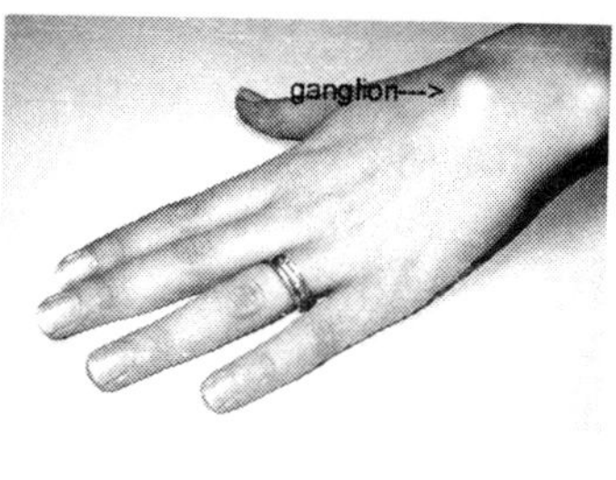

- **gap junction**

1. channel formed by proteins that allows ions and other molecules to pass between adjacent cells.
2. a narrow gap between plasma membranes of two animal cells, spanned by protein channels. They allow chemical substances or electrical signals to pass from cell to cell.

- **gape**

flehmen response in male cats.

- **gaping**

young song birds react to tactile stimuli by stretching their neck and opening the beak vertically upward.

- **gastrula**

a two-layered, later three-layered, animal embryonic stage.

- **gastrulation**

the process by which a blastula develops into a gastrula, usually by an involution of cells.

- **gene**

1. a unit of inherited material - a hereditary factor.

2. the functional unit of heredity; a cistron.
3. segment of DNA that specifies the structure of proteins (e.g., serum albumin); See also **Locus.**
4. the hereditary determinant of a specified characteristic of an individual; specific sequences of nucleotides in DNA.

- **gene amplification**
  any of the strategies that give rise to multiple copies of certain genes, thus facilitating the rapid synthesis of a product (such as rRNA for ribosomes) for which the demand is great.

- **gene cloning**
  formation by a bacterium, carrying foreign genes in a recombinant plasmid, of a clone of identical cells containing the replicated foreign genes.

- **gene conversion**
  when a gene is altered from within the genome. This is thought to occur through a breaking of both strands of DNA and the damaged or missing sequence on one chromosome is copied from the homologous strand and inserted into the damaged chromosomes.

- **gene delivery**
  this is a general term for the introduction of new genetic elements into the genomes of living cells. The delivery problem is essentially conditioned by the fact that the new genetic elements are usually large, and by the presence of the outer cell membrane and the nuclear membrane acting as barriers to incorporation of the new DNA into the genome already present in the nucleus. Viruses possess various natural biochemical methods for achieving gene delivery; artificial gene delivery is one of the essential problems of 'genetic engineering'. The most important barrier is apparently the outer cell membrane, which is essentially a lipid barrier, and introduction of any large complex into the cell requires a fusion of one kind or another with this membrane. Liposomes, which consist of lipid membranes themselves, and which can fuse with outer cell membranes, are thus

potential vehicles for delivery of many substances, including DNA.

- **gene family**
two or more loci with similar nucleotide sequences, that have been derived from a common ancestral sequence.

- **gene flow**
1. the consequence of cross-fertilisation between members of a species across boundaries between populations, or within populations, which results in the spread of genes across and between populations.
2. the movement of genes within a population or from one population to another, resulting from the dispersal of gametes (such as pollen by wind) or of zygotes (as by migration of fertile females, for example.)
3. the movement of genes from one part of a population to another, or from one population to another, via gametes.

- **gene pool**
1. the totality of the genes of a given sexual population existing at a particular time.
2. the sum total of all the genes of all the individuals in a population.
3. the sum total of all the genes that exist among all the individuals of a species.

- **gene regulation**
any of the strategies by which the rate of expression of a gene can be regulated, as by controlling the rate of transcription.

- **genealogy**
the study of the development of plants and animals from earlier forms.

- **generalist**
feeds on a wide range of food types; a jack-of-all-trades.

- **generalisation**
see **stimulus generalisation or response generalisation.**

- **generalised**
applying to an organism or character judged to be unchanged from a presumed common ancestral state.

- **generation**
the individuals of a species which are separated from a common ancestor by the same number of broods in the direct line of descent.

- **generation length**
not quite the average length of time between the birth of a parent and the birth of its offspring (the 'cohort generation time'): the production by offspring of their own offspring during the life of the original parent makes the average, computed over many generations, less than the cohort generation time.

- **generation time**
1. the average duration of a life cycle between birth and reproduction.
2. the mean period of time between reproduction of the parent generation and reproduction of the first filial generation.
3. the average age when females of a population begin reproducing, which greatly influences the intrinsic rate of increase.

- **generational mutualism**
a mutualistic relationship between members of different generations, as between parents and offspring.

- **genet**
the organism developed from a zygote. The term is used especially for modular organisms and members of a clone to define the *genetic individual* and to contrast with 'ramet' the potentially physiologically independent part that may arise from the iterative process by which modular organisms grow.

- **genetic code**
the code that relates nucleotide sequences in nucleic acids to amino-acid sequences. Each triplet of nucleotides designates a particular amino acid; thus, the genetic code allows the translation of information stored in DNA and the use of that information in protein synthesis.

- **genetic colonisation**
a novel sort of parasitism practiced by agrobacteria with plants; introduction of genetic information into a host, which induces the host to synthesise products that only the inducer can use.

- **genetic death**
a death due to genotype, hence contributing to natural selection. Deaths that are random with respect to

genotype are not genetic deaths.

- **genetic distance**
any of several measures of the degree of genetic difference between populations, based on differences in allele frequencies.

- **genetic engineering**
1. any change in the genetic constitution of an organism brought about by artificial means other than simple artificial selection and which would not usually occur in nature, such as the introduction of a gene from one species to another.
2. the artificial transfer of specific genes from one organism to another.

- **genetic equilibrium**
a state reached by a population when gene and genotype frequencies do not change in successive generations.

- **genetic evolution**
changes in gene frequency over generations.

- **genetic load**
the difference between the average fitness of a population and its maximal possible fitness, that of the fittest genotype (whose fitness is equal to 1); the sum total of genetic deaths per generation (as a proportion of the population).

- **genetic marker**
a gene mutation that has phenotypic effects useful for tracing the chromosome on which it is located.

- **genetic mosaic**
an individual whose tissues are a mix of different genotypes.

- **gestalt psychology**
a branch of psychology that attributes behavioural responses to the recognition of the sum of all elements of a configuration (in contrast to simple cues).

- **ghost of competition past**
the term stresses that interspecific competition, acting as an evolutionary force in the past, has often left its mark on the behaviour, distribution or morphology of species, even when there is no present-day competition between them.

- **giant fibre or giant neuron**
  unusually large cells that are found in the nervous systems of many invertebrates and some vertebrates that are used to trigger escape or startle responses.

- **giant motor neuron**
  motor Giant; Exceptionally large motor neuron of the crayfish which sends an axon branch to every fast flexor muscle fibre on both sides of each abdominal segment; Synonyms: Motor Giant.

- **giant neuron**
  neuron in an invertebrate in which the axon diameter is especially large, often more than 10 times larger than other axons in the nervous system.

- **gibberellin**
  a plant hormone which affects various phenomena, including stem elongation.

- **gill**
  an evaginated area of the body wall of an animal. The structure is specialised for gas exchange.

- **GIS (Geographic Information Systems)**
  GIS is both a database designed to handle geographic data as well as a set of computer operations that can be used to analyse the data. In a sense, GIS can be thought of as a higher order map.

- **gizzard**
  a chamber of an animal's digestive tract specialised for grinding food.

- **glial cell**
  1. non-neural cells in the nervous system.
  2. a nonconducting cell of the nervous system that provides support, insulation, and protection for the neurons.

- **global stability**
  the tendency of a community to return to its original state even when subjected to a large perturbation.

- **global warming**
  1. the process by which the Earth's atmosphere is warming because of the release of 'greenhouse gases,' such as carbon dioxide. These gases are released into the air from burning gas, oil, coal, wood,

and other resources and trap heat in an action similar to that of the walls of a greenhouse.

2. the term given to the possibility that Earth's atmosphere is gradually warming because of the greenhouse effect of carbon dioxide and other gases. Global warming is thought by many to be the most serious global environmental issue facing our society.

3. the predicted warming of the planet resulting from increasing atmospheric concentrations of radiative gases such as carbon dioxide, methane, nitrous oxide and chlorofluorocarbons.

- **glomerulus**

1. discrete, globular tangle of densely packed dendrites and axons found in the vertebrate olfactory bulb and in invertebrate olfactory processing centres.

2. a capillary bed within Bowman's capsule of the nephron; the site of ultrafiltration.

- **glucose**

a six carbon sugar which plays a central role in cellular metabolism.

- **glycocalyx**

the layer of protein and carbohydrates just outside the plasma membrane of an animal cell; in general, the proteins are anchored in the membrane and the carbohydrates are bound to the proteins.

- **glycogen**

a long, branched polymer of glucose subunits that is stored in the muscles and liver of animals and is metabolised as a source of energy.

- **glycolysis**

a metabolic pathway which occurs in the cytoplasm of cells and during which glucose is oxidised anaerobically to form pyruvic acid.

- **glycoprotein**

a protein with covalently linked sugar residues. The sugars may be bound to OH side chains of the polypeptide (O-linked) or the amide nitrogen of asparagine side chains (N-linked).

- **glycoside**

a derivative of glucose (or another sugar) in which one

hydrogen atom is replaced by an organic radical.

- **glycosidic**
  a type of bond which links monosaccharide subunits together in (di- or polysaccharides.)

- **glyoxysome**
  a type of microbody found in plants, in which stored lipids are converted to carbohydrates.

- **gobbling**
  vocalisation of mature male turkeys produced usually as a series of 3-7 sounds accompanied in most cases by head movement. [HWS]

- **golgi apparatus**
  a system of concentrically folded membranes found in the cytoplasm of eukaryotic cells. Plays a role in the production and release of secretory materials such as the digestive enzymes manufactured in the pancreas.

- **gonad**
  1. organs of animals which produce gametes.
  2. the site of production of gametes (testis, where sperm are produced, or ovary, where eggs are produced.)

- **gonadotropin**
  refers to a member of a group of hormones capable of promoting growth and function of the gonads. Includes hormones such as follicle stimulating hormone (FSH) and luteinising hormone (LH) which are stimulatory to the gonads.

- **gondwanaland**
  in the late Carboniferous the land masses were divided into two supercontinents, Laurasia to the north and Gondwana to the south.

- **good gene's hypothesis**
  1. the argument that mate choice advances direct fitness because it enhances the survival chances of offspring of choosy individuals by providing them with genes that promote viability.
  2. refers to an argument explaining female choice in sexual selection where female's mate with males who have genotypes that will be more likely to produce successful offspring.

- **grade**
  a level of phenotypic organisation attained by one or more species during evolution.

- **graded potential**
  changes in electric potential on the membrane of living cells which do not lead to action potentials.

- **graded response**
  a response that varies in preparation to the strength of the stimulus evoking it.

- **gradient analysis**
  the analysis of species composition along a gradient of environmental conditions.

- **gradual evolution**
  evolutionary change occurs slowly and at a more or less constant rate

- **gradualism**
  a view of Earth's history that attributes profound change to the cumulative product of slow but continuous processes.

- **granivores**
  animals that eat seeds.

- **granum**
  a stack like grouping of photosynthetic membranes in a chloroplast.

- **grasslands**
  areas of abundant grasses and shrubs with very few trees, like prairies or meadows. Grasslands can be found in many places. Large temperate grasslands occur naturally in central North America (prairies), southern South America (pampas), central Asia (steppes), southern Africa (savannas) and Australia.

- **grass-pulling**
  when two herring gulls happen to meet at their territorial boundary, a confrontation often occurs. The opponents viciously pull up tufts of grass. The behaviour has been explained as a conflict between two mutually inconsistent drives, flight and fight. This results in a displacement activity, in this case the grass pulling was thought to resemble nest building. However on closer analysis the behaviour is more like the movements used during fighting and so should be considered a redirected behaviour.

- **gravid**
  female carrying eggs or young.

- **gravitropism**

  a response of a plant or animal in response to gravity.

- **grazer**

  a consumer which attacks large numbers of large prey during its lifetime, but removes only a part of each prey individual, so that the effect, although often harmful, is rarely lethal in the short term and never predictably lethal.

- **grazer - scrapers**

  aquatic animals that graze the organic layer of algae, micro organisms and dead organic matter on stones and other substrates.

- **grazing**

  feeding on herbage, algae or phytoplankton by consuming the whole food plant or by cropping the entire surface growth.

- **green revolution**

  refers to the development and introduction of new varieties of wheat and rice (mainly) that increased yields per acre dramatically in some countries.

- **greenhouse effect**

  1. warming of the earth's atmosphere as a result of increases in CO_2 and other gases.
  2. the trapping of heat in the Earth's atmosphere by certain gases such as carbon dioxide, methane, and nitrous oxide. Some scientists predict that the temperature and sea level rise associated with global warming could adversely affect biodiversity.
  3. the warming of the Earth due to atmospheric accumulation of carbon dioxide which absorbs infrared radiation and slows its escape from the irradiated Earth.
  4. a natural effect that traps heat in the atmosphere (troposphere) near the earth's surface. Some of the heat flowing back toward space from the earth's surface is absorbed by water

vapour, carbon dioxide, ozone, and several other gases in the lower atmosphere (troposphere) and then radiated back toward the earth's surface. If the atmospheric concentrations of these greenhouse gases rise and are not removed by other natural processes, the average temperature of the lower atmosphere will gradually increase.

- **greenhouse gases**
  gases in the earth's lower atmosphere (troposphere) that cause the greenhouse effect. Examples are carbon dioxide, chlorofluorocarbons, ozone, methane, water vapour and nitrous oxide.

- **gregariousness**
  the tendency of animals to form social groups.

- **Gregor Mendel**
  the first to make quantitative observations of the patterns of inheritance and proposing plausible explanations for them.

- **grey water**
  all used water other than that originating from toilets and bidets.

- **grooming**
  behavior associated with care for the body surface.

- **Gross Primary Production (GPP)**
  the total fixation of energy by photosynthesis in a region.

- **gross primary productivity**
  the rate at which an ecosystem's producers capture and store a given amount of chemical energy as biomass in a given length of time.

- **ground fire**
  a fire that burns along the forest floor and does not af-

fect trees with thick bark or high crowns.

- **ground tissue**

a tissue of mostly parenchyma cells that makes up the bulk of a young plant and fills the space between the dermal and vascular tissue systems.

- **ground water**

the supply of fresh water under the earth's surface in an aquifer or in the soil.

- **ground-scratching**

a sequence of rapid foot movements performed that manipulate the ground surface in poultry.

- **group selection**

1. the evolutionary process which is supposed to act through the different numbers of descendants left by groups rather than by individuals.
2. the process of selection in which a group of individuals of one deme rather than a single individual is the target of selection; a process by which characters can be selected which benefit the group but not the individual (interdemic selection). The differential rate of origination or extinction of whole populations (or species, if the term is used broadly) on the basis of differences among them in one or more characteristics.
3. a method of tree harvest in which trees are removed periodically in small groups. This silvicultural treatment results in small openings that form mosaics of age class groups in the forest.

- **group territory**

a territory held by a group of three or more individuals. Group territories are normally but not invariably held all or most of the year. In most cases they are all-purpose territories, although. Some colonial bee-eaters are exceptions.

- **group-living**

a species may be termed group-living if at any time in

the year it normally lives in groups. Group-living, therefore, encompasses not only the communally breeding species but also species that breed colonially without helpers, those that breed in isolated pairs but form groups when not breeding, and any species 'that for whatever reason forms social groups on occasion.

- **growth factor**
  a protein that must be present in a cell's environment for its normal growth and development.

- **guano**
  manure from animals like birds and bats.

- **guard cell**
  a specialised epidermal cell that regulates the size of stoma of a leaf.

- **guild**
  a group of species that exploit the same class of environmental resources in a similar way.

- **gular fluttering**
  a thermoregulatory behaviour of some birds analogous to panting in which air is moved rapidly over the west surfaces on the inside of the mouth. Refers specifically to the associated behaviour where the skin on the throat flutters rapidly.

- **gymnosperm**
  a vascular plant that bears naked seeds not enclosed in any specialised chambers.

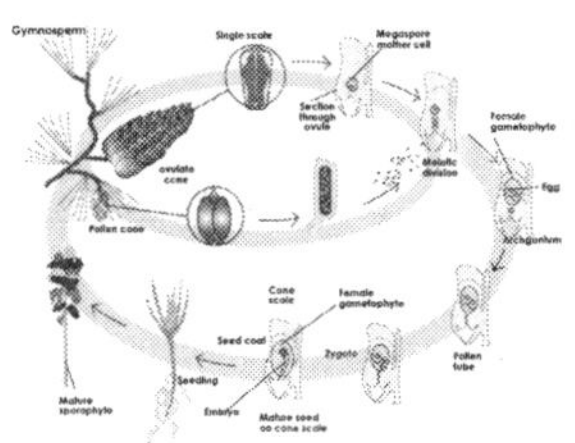

- **h²**
  see **heritability**

- **habit**
  in biology, the characteristic form or mode of growth of an organism.

- **habitat**
  1. place where a micro organism, plant or animal lives.
  2. any part of the biosphere where a species can successfully live and reproduce.
  3. the area where a plant or animal lives and grows under natural conditions.
  4. the area in which an animal, plant, or micro organism lives and finds the nu-

trients, water, sunlight, shelter, living space, and other essentials it needs to survive. Habitat loss, which includes the destruction, degradation and fragmentation of habitats, is the primary cause of biodiversity loss.
5. the kind of place where a given organism normally lives.

- **habitat capability**
the ability of a land area or plant community to support a given species of wildlife.

- **habitat diversity**
1. the range of habitats present in a region.
2. a number of different types of wildlife habitat within a given area.

- **habitat diversity index**
a measure of improvement in habitat diversity.

- **habitat saturation**
the condition that occurs when that point is reached on the continuum of competitor pressure where all suitable territories are occupied and defended. Competitor pressure refers to the number of individuals seeking territories. If territories are essentially incompressible, saturation is a clearly defined point.

- **habitat selection**
the capacity of an organism (usually an animal) to choose a habitat in which to perform activities.

- **habitat type**
a way to classify land area. A habitat type can support certain climax vegetation, both tree and undergrowth species. Habitat typing can indicate the biological potential of a site.

- **habitat**
place or type of place where an organism, population or community lives.

- **habituation**
1. the process that results in a long-lasting decline in the receptiveness of interneurons to the input from sensory neurons or other interneurons.

2. the decrement in response-strength which occurs with the repeated elicitation of a response. Response decrement is a function of the number of elicitations, it is specific to the response (and independent of the stimuli eliciting it) and from which recovery is slow or absent until the stimulus is increased.

3. gradual decline in response to recurrent stimuli without specific significance. The tendency to become familiar with a stimulus of no consequence after repeated exposure to it. Learned suppression of response to a repeated stimulus.

- **hair cell**

a common type of mechanoreceptor in the vertebrate ear, in the lateral line system of fishes and amphibians and in arthropod statocsts, that detects motion of various kinds.

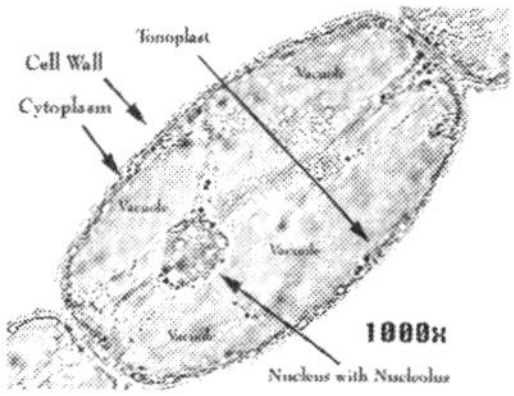

- **haldane's rule**

the heterogametic sex in species hybrids is often wholly or partially sterile owing to imbalance between the single sex-chromosome derived from one parent species.

- **halophyte**

a plant that tolerates very salty soil.

- **hawking**

feeding in flight, particularly leaving a perch, catching a prey and returning to the perch.

- **hazardous waste**

any solid, liquid, or containerised gas that can catch fire easily, is corrosive to skin tissue or metals, is unstable and can explode or release toxic fumes, or has harmful concentrations of one or more toxic materials that can leach out.

- **head flagging**

an appeasement gesture during the mating and pair formation ceremony in the black-headed gull which consists of turning away the black facial mask from the opponent or mate.

- **heat**
  the total amount of kinetic energy due to molecular motion in a body of matter. Heat is energy in its most random form.

- **heat of vaporisation**
  the amount of energy absorbed by a substance when it changes state to a gas. Water absorbs approximately 580 calories per gram when it changes from liquid water to water vapour.

- **helper**
  the situation where adult individuals are assisting with the rearing of young that are not their own although they are usually close relatives.

- **helper t cell**
  a type of T cell that is required by some B cells to help them make antibodies or that helps other T cells respond to antigens or secrete lymphokines or interleukins.

- **helping behaviour**
  an individual that performs parent-like behaviour toward young that are not genetically its own offspring, typically in company with the young's real parents (thus excluding brood parasitism and brood capture). Helpers may be breeders or nonbreeders. They may or may not benefit the young or their parents. Helpers may be altruistic, cooperative or selfish. Note that breeding status and conferral of benefit or harm to recipient or helper are irrelevant to the definition but helpers are often collateral kin.

- **hemagglutinin**
  a surface antigen on influenza viruses which controls infectivity by associating with receptors on host erythrocytes or other cells.

- **hematopoiesis**
  the formation of blood.

- **hemoglobin**
  an iron-containing respiratory pigment found in many organisms.

- **hemolymph**
  in invertebrates with open circulatory systems, the body fluid that bathes tissues.

- **hemophilia**
a genetic disease resulting from an abnormal sex-linked recessive gene, characterised by excessive bleeding following injury.

- **hepatic**
pertaining to the liver.

- **heterozygous**
the condition whereby two different alleles of the gene are present within the same cell.

- **heterozygous advantage**
the manifestation of higher fitness by heterozygotes at a specific locus than by homozygotes.

- **heuristic**
providing direction in the solution of a problem but otherwise unjustified; a rule of thumb.

- **hibernaculum**
the domicile of an animal in which it hibernates or overwinters.

- **hibernate**
to remain dormant during the winter period.

- **hibernation**
a form of winter dormancy characterised by a slowing of metabolic processes and a marked fall in body temperature.

- **hiding area/cover**
vegetation capable of hiding 90% of an adult elk or deer from human's view at a distance of 200 feet or less.

- **hierarchy**
1. the occurrence of things in a certain rank or order, such as the hierarchical organi-sation of behaviour, hierarchical organisation of neural centres or hierarchical organisation of a social group.
Hierarchy refers to situations in which there is one entity-a boss - that controls all entities directly below it. In a hierarchy, information may flow from lower to higher levels, but control is directed from higher to lower levels only. That is, in a hierarchy the elements are ordered in such a way that higher-ranking elements control lower ones. A linear social hierarchy is formed as a straight line of dominance-subordinate relationships through the whole group. The highest

ranking animal dominates all, the second highest all but the first, the third highest all but the first two, etc.

- **hilum**
a region where blood vessels, nerves, ducts, enter an organ.

- **histamine**
a substance released by injured cells that causes blood vessels to dilate during an inflammatory response.

- **histology**
the study of tissues.

- **histone**
one of a class of proteins that are constituents of the chromosomes in eukaryotes.

- **hitchhiking**
the use of animals of other species as means of transport. Hitchhiking genes are silent genes that get into the gene pool by their association with other genes.

- **HIV-1**
acute human immunodeficiency virus type-1 is the subtype of HIV (human immune deficiency virus) that causes most cases of AIDS in the Western Hemisphere, Europe, and Central, South, and East Africa. HIV is a retrovirus (subclass lentivirus), and retroviruses are single-stranded RNA viruses that have an enzyme called reverse transcriptase. With this enzyme the viral RNA is used as a template to produce viral DNA from cellular material. This DNA is then incorporated into the host cell's genome, where it codes for the synthesis of viral components. An HIV-1 infection should be distinguished from AIDS. Acquired immunodeficiency syndrome (AIDS) is a secondary immunodeficiency syndrome resulting from HIV infection and characterised by opportunistic infections, malignancies, neurologic dysfunction, and a variety of other syndromes.

- **hoarding**
many birds and mammals gather food for the winter and hide it at certain places.

- **holism**
the belief that the sum of the parts is greater than the individual components, that we

cannot understand phenomena through reductionism.

- **holoblastic**
  a type of cleavage in which there is complete division of the egg, as in eggs having little yolk (sea urchin) or a moderate amount of yolk (frog).

- **holometabolous**
  complete metamorphosis in insects, egg-larva, pupae (contrast hemimetabolous).

- **holoparasite**
  parasitic plants which lack chlorophyll and are therefore wholly dependent on their host plant for the supply of water, nutrients and fixed carbon.

- **home range**
  1. the entire area which is normally occupied by an individual animal or stable group throughout the life of the animal. This area is distinct from a territory or individual distance.
  2. an area that an animal occupies but does not defend, in contrast to a territory, which is defended.
  3. an area within which an animal tends to confine all or nearly all its activities for a long period of time.

- **host range**
  the limited number of host species, tissues, or cells that a parasite (including viruses and bacteria) can infect.

- **huddling**
  contact behaviour.

- **human capital**
  physical and mental talents of people used to produce, distribute, and sell an economic good.

- **human ecology**
  see **ecology.**

- **human ethology**
  1. a special and recent branch of ethology which endeavours to examine human behaviour with the aid of methods that have been used in the study of animal behaviour from an ethological point of view.
  2. human ethology tries to elucidate the phylogenetic bases of human behaviour and to determine with the help of a comparative study of cultures, the innate components of these complex behaviour patterns.

- **human time scale**
occurring within a short enough time frame that the event can be perceived, remembered, and recounted by humans through oral traditions, written histories or other mechanisms of human memory.

- **humoral immunity**
the type of immunity that fights bacteria and viruses in body fluids with antibodies that circulate in blood plasma and lymph, fluids formerly called humors.

- **hunger**
the bodily sensation which appears under condition of food deprivation and which results in the desire and craving for food.

- **hunting**
an aspect of predatory behaviour involving an animal of one species, the predator, capturing a member of another species, the prey. May also refer to the active pursuit of prey after sighting.

- **hybrid**
1. a progeny resulting from hybridisation.
2. in evolutionary biology, a cross between two species. In genetics, a cross between two genetic types.

- **hybrid breakdown**
a hybrid is fully viable but is selected against because of its phenotype.

- **hybrid sterility**
the sterility of hybrids, usually a consequence of failure of meiosis.

- **hybrid vigour**
increased vitality (compared to that of either parent stock) in the hybrid offspring of two different, inbred parents.

- **hybrid**
the offspring of two parents from separate (though closely related) species. Usually sterile, though occasionally able to breed back into one of the parent lines. A hybrid can almost never produce viable offspring when mated with another hybrid. A common example is a mule, which is produced by breeding a horse with a donkey (note that the horse must be the mother, due to the large size of the foal). Hybridisation is fairly com-

mon among wind-pollinated plants, while hybridisation is quite uncommon among higher animals.

- **hybridisation**

1. the crossing of individuals from two dissimilar populations or species.
2. the process whereby a hybrid results from interbreeding two species;
3. DNA hybridisation is the comparison of whole genomes of two species by estimating the extent of hydrogen bonding that occurs between single-stranded DNA obtained from the two species.

- **hybridoma**

a hybrid cell that produces monoclonal antibodies in culture, formed by the fusion of a myeloma cell with a normal antibody-producing lymphocyte.

- **hydrocarbon**

any compound made of only carbon and hydrogen.

- **hydrogen bond**

a type of bond formed when the partially positive hydrogen atom of a polar covalent bond in one molecule is attracted to the partially negative atom of a polar covalent bond in another.

- **hydrogen ion**

a single proton with a charge of +1. The dissociation of a water molecule (H_2O) leads to the generation of a hydroxide ion (OH^-) and a hydrogen ion (H^+).

- **hydrologic cycle**

1. also called the water cycle, this is the process of water evaporating, condensing, falling to the ground as precipitation, and returning to the ocean as run-off.
2. biogeochemical cycle that collects, purifies, and distributes the earth's fixed supply of water, from the environment to living organisms and then back to the environment.

- **hydrological cycle**

the movement of water from ocean, by evaporation, to atmosphere, to land and back, via river flow, to ocean.

- **hydrology**

the science dealing with the study of water on the surface of the land, in the soil and underlying rocks and in the atmosphere.

• **hydrolysis**
the chemical reaction that breaks a covalent bond through the addition of hydrogen (from a water molecule) to the atom forming one side of the original bond, and a hydroxyl group to the atom on the other side.

• **hydrophilia**
an overwhelming desire for water.

• **hydrophilic**
having an affinity for water.

• **hydrophobic**
the physicochemical property whereby a substance or region of a molecule resists association with water molecules.

• **hydrophobic interaction**
a type of weak chemical bond formed when molecules that do not mix with water coalesce to exclude the water.

• **hydrosphere**
the earth's liquid water (oceans, lakes other bodies of surface water and underground water), the earth's frozen water (polar ice caps, floating ice caps and ice in soil known as permafrost) and small amounts of water vapour in the atmosphere.

• **hypermorphosis**
exaggeration of the features of a descendant form compared to those of its ancestor due to an increase during evolution of the duration of ontogenetic development. Specifically peramorphosis due to an increase in the endpoint of development.

• **hyperosmotic**
a solution with a greater solute concentration than another, a hypoosmotic solution. If the two solutions are separated from one another by a membrane permeable to water, water would tend to move from the hypo- to the hyperosmotic side.

• **hyperphagia**
overeating.

• **hyperpolarisation**
an electrical state whereby the inside of the cell is made more negative relative to the outside than was the case at resting potential. A neuron membrane is hyperpolarised if the voltage is increased from the resting potential of about -70 mV,

reducing the chance that a nerve impulse will be transmitted.

- **hypertrophy**
  abnormal enlargement, excessive growth.

- **hypha**
  a fungal filament.

- **hypocotyl**
  the portion of the axis of a plant embryo below the point of attachment of the cotyledons; forms the base of the shoot and the root.

- **hypoosmotic solution**
  a solution with a lesser solute concentration than another, a hyperosmotic solution. If the two solutions are separated from one another by a membrane permeable to water, water would tend to move from the hypo- to the hyperosmotic side.

- **hypothalamus**
  1. the ventral part of the anterior division of the brain stem (forebrain, diencephalon) in vertebrates. It contains areas involved in the regulation of such functions as eating, drinking, and reproduction. For example, neurosecretory cells in the hypothalamus regulate the output of hormones from the pituitary gland and thus affect the gonadal, thyroid, and adrenocortical functions, as well as exerting more direct effects through production of prolactin, oxytocin and other hormonal substances.
  2. part of the posterior portion of the vertebrate forebrain. This structure contains several centres of the autonomic nervous system and centres of emotion.

- **hypothesis**
  1. an assertion or working explanation that leads to testable predictions: an assumption providing an explanation of observed facts, proposed in order to test its consequences.
  2. a formal statement of supposition offered to explain observations. Note that a hypothesis is only useful if it can be tested. Even if correct, it is not scientifically useful if untestable.

- **hypothetico-deductive**
  a method used to test hypotheses. If deductions formulated from the hypoth-

esis are tested and proven false, the hypothesis is rejected.

- **ichnospecies**
  species which are defined on the basis of trace fossils only.

- **iconic display**
  signals that resemble their object or meaning.

- **ideal free distribution**
  1. competitors adjust their distribution in relation to habitat quality so that each individual enjoys the same rate of acquisition of resources. This assumes that animals are free to go where they will do best and that the animals are ideal in having complete information about the availability of resources.
  2. theoretical expectation in which individuals should distribute themselves among habitats so that every individual maximises its net rate of return.

- **identical by descent**
  describes copies of an allele that may be traced back through an arbitrary number of generations without mutation to a common ancestor of the organisms that carry the copies.

- **idiopathic**
  pathologic state of unknown cause.

- **idiothetic orientation**
  orientation in the absence of external stimuli such as the ability of an animal to orient in space without using external cues.

- **IEC**
  International Ethological Conference.

- **igneous rock**
  rocks formed when high temperature, molten mineral matter cooled and solidified.

- **imaginal disk**
  an island of undifferentiated cells in an insect larva, which are committed (determined) to form a particular organ during metamorphosis to the adult.

- **imbibition**
  the soaking of water into a porous material that is hydrophilic.

- **imitative behaviour**
  an animal is placed in the environment of another animal. If the first animal then makes some response that is not species-specific, and thereafter the second animal makes the same response under conditions when it can be demonstrated that the response of the first animal is the stimulus for the response of the second, then the second animal is said to imitate the first and the behaviour is called imitative.

- **immigration**
  entry of organisms to a population from elsewhere.

- **immobility**
  tonic immobility (sometimes referred to as hypnosis) is a defensive freezing response of animals when faced with a predator or other traumatic event. This is an important aspect of their camouflage.

- **immune response**
  1. a primary immune response is the initial response to an antigen, which appears after a lag of a few days.
  2. a secondary immune response is the response elicited when the animal encounters the same antigen at a later time. The secondary response is normally more rapid, of greater magnitude and of longer duration than the primary response.
  3. a response mounted by an animal in which specific antibodies (vertebrates only) and/or cytotoxic cells are produced against invading microorganisms, parasites, transplanted tissue and many other substances which are recognised as foreign by the body.

- **immunoglobulins**
  the class of proteins comprising the antibodies.

- **immunological**
  1. immunological distance is the amount of difference between two proteins as measured by the strength of the antigen-antibody reaction between them.
  2. immunological tolerance is a mechanism by

which an animal does not mount an immune response to the antigenic determinants of its own macromolecules.

- **immunological distance**
  the cross reactions of antibodies to the proteins of a specific serum. If antisera are not produced then the protein is considered more distantly related than if there is a weak or strong reaction. (also called antigenic distance.)

- **impaling**
  behavior shown by shrikes and a few other species in which they attach food to sharp objects in the environment such as thorns or barbed wire.

- **imprinting**
  capacity to learn specific types of information at certain critical periods in development.

- **in vitro fertilisation**
  fertilisation in laboratory containers followed by artificial implantation of the early embryo in the mother's uterus.

- **inbreeding**
  preferential mating between relatives, extreme inbreeding: mating between sibs, half-sibs, parent-offspring.

- **inbreeding coefficient**
  the probability that an allele present in a common ancestor is homozygous in the inbred individuals.

- **inbreeding depression**
  a loss of vigour amongst offspring occurring when closely related individuals mate, resulting from the expression of numbers of deleterious genes in a homozygous state and from a generally low level of heterozygosity.

- **incentive**
  a type of motivation that derives from expectation of reward or punishment. When an animal has learning a particular task to obtain a food reward, for example, it will perform more strongly when its expectation of reward is high.

- **incest avoidance**
  a general term for mechanisms that obstruct mating between closely related in-

dividuals and so prevents the negative consequences of inbreeding.

- **incidence function**
  the relationship between island size and the proportion of islands of that size occupied by the species in question.
- **inciting**
  the most widely distributed courtship behaviour in female ducks. The female orients towards her partner or her prospective mate, follows him and threatens at the same time towards another conspecific male. In communally courting species this behaviour serves to isolate a male from the group and stimulate him to become aggressive against other conspecific males. The behaviour is also found in already established pairs.
- **inclusive fitness**
  the sum of an individual's own fitness (direct components) plus all its influence on fitness in its relatives (indirect component) other than direct descendants; hence the total effect of kin selection with reference to an individual.
- **incomplete dominance**
  a type of inheritance in which F1 hybrids have an appearance that is intermediate between the phenotypes of the parental varieties.
- **incomplete flower**
  a flower lacking sepals, petals, stamens or carpels.
- **incomplete metamorphosis**
  a type of development in certain insects, such as grasshoppers, in which the larvae resemble adults but are smaller and have different body proportions. The animal goes through a series of molts, each time looking more like an adult, until it reaches full size.
- **incubation**
  of birds, sitting on eggs.
- **incubation period**
  the period during which the embryo in an egg develops before hatching. Also, in epidemiology, the period between infection and the appearance of symptoms of disease.
- **independent variable**
  in an experiment, the condition manipulated by the experimenter who mea-

sures its effect on the dependent variable.

- **indeterminate**
  1. a type of cleavage exhibited during the embryonic development in deuterostomes, in which each cell produced by early cleavage divisions retains the capacity to develop into a complete embryo;
  2. a type of growth exhibited by plants- they continue to grow as long as they live, because they always retain meristematic cells capable of undergoing mitosis.

- **indexic display**
  signals that point out their objects, e.g., the stickleback nest presentation display.

- **indicator species**
  1. a plant or animal species related to a particular kind of environment. Its presence indicates that specific habitat conditions are also present.
  2. species that serve as early warnings that a community or an ecosystem is being degraded. Fish and amphibians make particularly excellent indicator species. Large predators (those generally at the apex of the food pyramid) are also good indicators in many habitats.

- **indigenous (species).**
  any species of wildlife native to a given land or water area by natural occurrence.

- **indigenous custodians**
  indigenous people with cultural responsibility for managing their traditional land and sea, whether or not they are currently the statutory owners or managers of those areas.

- **indigenous peoples**
  Australians who identify as belonging to aboriginal or Torres Strait Islander groups and who are recognised by their communities as belonging to those groups.

- **indigenous vegetation**
  a species of plant that is naturally found in an area, either historically or currently and forms part of the natural biodiversity of the area. Ideally, plants used to revegetate an area should be grown from the cuttings or seed from indigenous plants taken from that area

or an area nearby to ensure that the same species are reintroduced, and are genetically similar to those originally in the area.

- **indirect fitness**
  the long-term reproductive success of an individual through their kin other than offspring. The component of inclusive fitness that arises from effects on non descendent kin and other non descendent co-gene-carriers. Direct effects are attributed to individuals who physically transmit their genes through their own gametes, usually to the affected individual; indirect effects are not mediated in this way and do not require the production or transmission of gametes by the individual.

- **indirect head scratching**
  the scratching pattern in birds in which the wing is dropped and the birds raises the ipsilateral leg over and above it to scratch the head. In contrast to direct head scratching.

- **indirect selection**
  the process that occurs when individuals differ in their effects on the survival of nondescendant kin, creating differences in the indirect fitness of the individuals interacting with this category of kin.

- **indirect spermatophore transmission**
  a number of terrestrial animals have evolved ways to effect fertilisation without copulation. Lacking copulatory organs, the males of these species convey spermatophore indirectly by depositing them on the ground and then inducing the female to pick them up in such a way as to achieve fertilisation. This mode of sperm transmission has evolved independently many times in animals living on relatively moist substrates, where desiccation of the spermatophore is unlikely. In vertebrates, it occurs in some species of salamanders, and among invertebrates, in many arthropods, including myriapods, scorpions, whip scorpions, spiders, mites and apter-ygote insects.

- **individual distance**
  the minimum distance tolerated between individuals

under normal social conditions; this distance is characteristic of the species but may show some variation depending on circumstances.

- **individual recognition**
  personal recognition between individuals.

- **individual selection**
  natural selection at the level of the individual where individual fitness is maximized. A form of natural selection consisting of non-random differences among different genotypes within a population in their contribution to subsequent generations.

- **individualistic concept**
  concept of the community as an association of species that occur together simply because of similarities in requirements and not as a result of a long co evolutionary history.

- **induced fit**
  the change in shape of the active site of an enzyme so that it binds more snugly to the substrate, induced by entry of the substrate.

- **induction**
  1. the ability of one group of embryonic cells to influence the development of another.
  2. a method in logic which proceeds from the specific to general and develops a general statement which explains all of the observations. Commonly used to formulate scientific hypotheses.

- **industrial melanism**
  a dark morph resulting from selection, where the selective pressure is associated with industrialization and the associated blackening of the environment with soot and smoke.

- **infanticide**
  the killing of infants.

- **infantilism**
  a return to juvenile behaviour patterns by adult animals especially in connection with courtship or submissive gestures.

- **infectious**
  1. an infectious disease is a disease caused by an infectious microbial or parasitic agent.

2. infectious hepatitis is the former name for hepatitis A.
3. infectious mononucleosis is an acute disease that affects many systems, caused by the Epstein-Barr virus.

- **inflammatory response**
  a line of defence triggered by penetration of the skin or mucous membranes, in which small blood vessels in the vicinity of an injury dilate and become leakier, enhancing infiltration of leukocytes; may also be widespread in the body.

- **inflorescence**
  the organ bearing an aggregation of flowers in a flowering plant.

- **information centres**
  an hypothesis explaining the formation of social groups in which group members derive information about the location of food or limited resources by participating in this group.

- **infostering**
  a part of an experimental design in which young are reared by individuals of the same strain but not their own parents.

- **infrared or IR light**
  the wavelength of electromagnetic radiation that is associated with our sensation of heat.

- **infrasound**
  sounds at frequencies below the range of human hearing.

- **ingestion**
  a heterotrophic mode of nutrition in which other organisms or detritus are eaten whole or in pieces.

- **ingestive behaviour**
  eating including suckling and drinking.

- **inhibition**
  1. the suppression of a condition or state; the hindrance or interruption of a process. Ethological inhibition refers to blockage of a behaviour pattern through certain endogenous or allochthonous stimuli.
  2. a hypothetical state of the animal sometimes used to account for decrements in response magnitude, or habit-strength, or both.
  3. a cortical process opposite to excitation and postulated as having the property of suppressing excitation

and hence, of suppressing the behaviour dependent on that excitation.
4. a block.

- **inhibitory postsynaptic potential**
  an electrical charge (hyperpolarisation) in the membrane of a postsynaptic neuron caused by the binding of an inhibitory neurotransmitter from a presynaptic cell to a postsynaptic receptor.

- **initiation codon**
  code (generally AUG) for the first amino acid in protein sequences which is generally formylmethionine.

- **innate**
  much of animal behaviour is innate in the sense that it inevitably appears as part of an animal's repertoire under normal and natural conditions. Genetic factors are necessary for such behavi-our but the species-characteristic processes of maturation and learning may be just as important. The naive dichotomy between nature and nurture which has characterised so much of the debate about innate behaviour bears no relation to the complex and subtle interactions between the various processes that are responsible for the development of behaviour patterns.
  1. refers to behaviour not learned or acquired during ontogeny; behaviour that is present at birth but not necessarily functional, or behaviour that may appear only in the course of ontogenetic maturation.
  2. a term applied to differences in genetic characters between two members of the same species that have been raised in the same environment.

- **Innate Releasing Mechanism (IRM)**
  all those hypothetical structures of the CNS of an organism which have an important part in eliciting the reaction, but does not include the motor system. An IRM may either originate through ontogeny or phylogeny. A hypothetical physiological neurosensory mechanism invoked to explain the action of sign stimuli.

- **inner cell mass**
a cluster of cells in a mammalian blastocyst that protrudes into one end of the cavity and subsequently develops into the embryo proper and some of the extraembryonic membranes.

- **innovation**
inventive or new behavio -ural combination.

- **inoculation**
the introduction of a disease organism or vaccine, usually through a deliberately made wound.

- **inositol triphosphate**
the second messenger, which functions as an intermediate between certain nonsteroid hormones and the third messenger, a rise in cytoplasmic Ca++ concentration.

- **input resistance**
tendency of a cell to impede the flow of current into the cell. Equal to half the square root of the membrane resistance times the internal resistance.

- **inquilinism**
interspecific relation in which a socially parasitic species of an insect spends its entire life cycle in the nests of its host species. Workers either are lacking or if present are usually scarce and degenerate in behaviour.

- **insemination**
introduction of sperm into reproductive tract of female.

- **insertion**
a mutation involving the addition of one or more nucleotide pairs to a gene.

- **insight**
ethologists speak of insight when an individual is able to master a new situation or task without the help of pre-programmed innate responses and without previously having attempted the solution. A gross difference in behaviour between two successive occasions on which behavior can occur, when the behaviour shown on the second occasion is close to what the experimenter has previously decided upon as a good or efficient solution.

- **insight learning**
  the ability of an animal to perform a correct or appropriate behaviour on the first attempt in a situation with which it has had no prior experience.

- **instantaneous speciation**
  the production of a single individual that is reproductively isolated from the species to which the parents belong and is reproductively capable of establishing a new species population.

- **instinct**
  capacity of an animal to complete a complex behaviour in response to a first-time encounter with a given situation; purposeful action without foresight of the ends and without previous education in the performance, an inborn behavioural mechanism that manifest itself in an ordered movement.

- **instinctive behaviour**
  a behaviour sequence involving inherited pathways of the central nervous system. Such sequences are no more subject to modification than are morphological traits of the individual.

- **instream flow**
  the quantity of water necessary to meet seasonal stream flow requirements to accomplish the purposes of the National Forests, including, but not limited to fisheries, visual quality and recreational opportunities.

- **instrumental conditioning**
  1. operant conditioning.
  2. a term applied to operant conditioning when the animal is not free to give the response except when the experimenter chooses.

- **insulin**
  the vertebrate hormone that lowers blood sugar levels by promoting the uptake of glucose by most body cells and promoting the synthesis and storage of glycogen in the liver; also stimulates protein and fat synthesis; secreted by endocrine cells of the pancreas called islets of Langerhans.

- **integral protein**
  a protein of biological membranes that penetrates into or spans the membrane.

- **integrated local area planning**
  a coordinated approach to providing services and infra-

structure. It aims to develop partnerships for effective long-term planning and resource allocation.

- **Integrated Pest Management**
IPM evaluates alternatives for managing forest pest populations, based on consideration of pest-host relationships.

- **integration**
process by which signals in the nervous system interact; the decision-making capability of neurons.

- **integration and excision**
integration: a recombination in which a genetic element is inserted; excision: genetic element deleted.

- **intellectual property**
intellectual property represents the property of your mind or intellect. This includes information people have as part of their cultural heritage, for example, knowledge about bush foods or oral history.

- **intelligence**
1. the ability of animals to learn.
2. the ability of animals to use insight learning.

- **intensity of abundance**
the number of individuals per habitable site in a community.

- **intensity of infection**
the number of parasites per host in a population.

- **intensity of response**
with a constant stimulus but under increasing values of variables involved in drive-establishing operations, or vice versa, a response may not only increase in magnitude but may alter in topography as well. Positions along such a graded continuum are referred to as different intensities of response.

- **intention movement**
behaviour patterns that precede, signal or prepare for another behaviour.

- **interbreed**
to breed with another kind or species; hybridise.

- **interdisciplinary team**
a team of individuals with skills from different disciplines that focuses on the same task or project.

- **interference competition**
competition between two organisms in which one

physically excludes the other from a portion of habitat and hence from the resources that could be exploited there.

- **interferon**
  a chemical messenger of the immune system, produced by virus-infected cells and capable of helping other cells resist the virus.

- **interglacial period**
  period between glaciations, during which species begin to recolonise locations occupied previously.

- **interleukin-1**
  a chemical regulator (cytokine) secreted by macrophages that have ingested a pathogen or foreign molecule and have bound with a helper T cell; stimulates T cells to grow and divide and elevates body temperature. Interleukin-2, secreted by activated T cells, stimulates helper T cells to proliferate more rapidly.

- **intermediate cut**
  the removal of trees from a stand sometime between the beginning or formation of the stand and the regeneration cut. Types of intermediate cuts include thinning, release and improvement cuttings.

- **intermediate filament**
  a component of the cytoskeleton that includes all filaments intermediate in size between microtubules and microfilaments.

- **intermittent stream**
  a stream that flows only at certain times of the year when it receives water from streams or from some surface source, such as melting snow.

- **interneuron**
  1. a neuron that connects two neurons, such as from a sensory neuron to a motor neuron.
  2. neuron that is neither purely sensory nor motor but connects and interfaces between other neurons.
  3. an association neuron; a nerve cell within the central nervous system that forms synapses with sensory and motor neurons and integrates sensory input and motor output.

- **internode**
  the segment of a plant stem between the points where leaves are attached.
- **interoceptor**
  a sensory receptor activated by stimuli occurring inside the body, e.g. proprioreceptors.
- **intersexual selection**
  female or male choice of mates.
- **ion**
  a charged chemical formed when an atom or group of atoms has more or less electrons than protons (rather than an equal number.)
- **ion channel**
  protein or group of proteins that form a pore through a cellular membrane allowing a particular class of ions to cross the membrane. Passive channels are always open, while active (or gated) channels are opened or closed by a signal. Examples of the latter include voltage-gated (open state controlled by changes in membrane potential) or ligand-gated (controlled by the binding of a chemical molecule) ion channels.
- **ionic bond**
  a chemical bond due to attraction between oppositely charged ions.
- **ionotropic receptor**
  ion channel that binds a neurotransmitter or other ligand causing a rapid change in its permeability to ions and thus a rapid change in membrane potential.
- **iridium**
  heavy element not normally present in the rock of the earth's crust but is common in meteors.
- **irrelevant behaviour**
  while the animal is performing one kind of behaviour it may suddenly break off and do something completely different and apparently not appropriate to the context. These are normally classified as displacement activities.
- **irretrievable**
  one of the categories of impacts mentioned in the National Environmental Policy Act to be included in statements of environmental impacts. An irretrievable effect

applies to losses of production or commitment of renewable natural resources. For example, while an area is used as a ski area, some or all of the timber production there is irretrievably lost. If the ski area closes, timber production could resume; the loss of timber production during the time that the area was devoted to winter sports is irretrievable. However, the loss of timber production during that time is not irreversible, because it is possible for timber production to resume if the area is no longer used as a ski area.

- **irreversible**
  a category of impacts mentioned in statements of environmental impacts that applies to non-renewable resources, such as minerals and archaeological sites. Irreversible effects can also refer to effects of actions that can be renewed only after a very long period of time, such as the loss of soil productivity.

- **irritability**
  the ability of all living organisms to react to influences from the external environment with a particular response.

- **irritable aggression**
  animals that are in pain or uncomfortable have a lowered threshold for aggression.

- **irruption**
  a rapid increase in population density often followed by a mass emigration.

- **island biogeography**
  the study of distribution of species and community composition on islands.

- **isocline**
  a line linking points, each of which gives rise to the same rate of population increase for the species being considered. The points may represent combinations of species densities or combinations of resource levels.

- **isogamy**
  1. condition in which the gametes from both partners in sexual reproduction are of equal size.
  2. a condition in which male and female gametes are morphologically indistinguishable.

3. male and female gametes are the same size.

- **isolating mechanisms**
any structural, physiological, behavioural or other feature of an individual, or any geographical or geological barrier, that prevents individuals of one population from successfully interbreeding with those of other populations.

- **isolation experiment**
also referred to as experience deprivation and deprivation experiment. Deliberate withholding of certain possibilities of experience in the rearing of an animal. With this method one can determine which capabilities develop normally, even in the absence of what might have appeared relevant experience, and what deficits go with what kinds of deprivation. Evidence from isolation experiments has been the basis of inferences about the components of a species' behaviour that are innate, but this is a controversial matter. Inference from an experiment with experience deprivation is possible in only one direction: if the behaviour of a deprived animal turns out to be normal, one can conclude that the experience withheld is not necessary to development of that behaviour. But if development is abnormal, the reverse conclusion does not necessarily follow, because the deficits could also be the result of unnatural features in the circumstances of the isolation experiment.

- **isomer**
molecules consisting of the same numbers and kinds of atoms, but differing in the way in which the atoms are combined.

- **isometry**
when the growth of one part of the body is exactly proportional to growth in another part of the body or to overall body size.

- **isosmotic**
solutions of equal concentration with respect to osmotic pressure.

- **isotherm**
  a line on a map that joins places having the same mean temperature.

- **isotonic**
  having the same osmotic pressure.

- **isotope**
  an atomic form of an element, containing a different number of neutrons than another isotope. Isotopes vary from one another with respect to atomic mass.

- **iteroparity**
  where organisms produce offspring in a series of separate events during and after each of which the organisms maintain themselves in a condition that favours survival to reproduce again subsequently (though the 'separate events' may merge into continuous reproduction.)

- **iteroparous**
  used of organisms that have repeated reproductive cycles.

- **joint nesting**
  a system of nesting in which two or more females of the same species normally lay their eggs in the same nest. In communally breeding species the joint-nesting females of the same species normally lay their eggs in the same nest. In communally breeding species, the joint-nesting females of the same species normally the care of the young (thus excluding egg parasitism).

- **jump**
  to spring from the ground or surface with the propulsive force being derived primarily from the action of the legs.

- **jumping genes**
  genes associated with transposable elements.

- **jurassic**
  a geological period of the Mesozoic era 210-140 my ago.

- **juvenile behaviour**
  the behaviour of young individuals. Like adult behaviour, juvenile behaviour is subject to natural selection and often differs from the behaviour of adults due to the different requirements of young individuals.

- **Juvenile Hormone (JH)**
  a hormone in arthropods secreted by the corpora al-

lata glands, which promotes the retention of larval characteristics.

- **Juxta Glomerular Apparatus (JGA)**
  specialized tissue located near the afferent arteriole that supplies blood to the kidney glomerulus; JGA raises blood pressure by producing renin, which activates angiotensin.

- **k**
  carrying capacity.

- **k selection**
  selection of life-history traits which promote an ability to make a large proportionate contribution to a population which stays close to its carrying capacity - the traits being, broadly, large size, delayed reproduction, iteroparity, a small reproductive allocation, much parental care, and the production of few but large offspring.

- **kairomone**
  any substance or blend of substances emitted by an organism that elicits a response adoptively favourable to the receiver but not to the emitter.

- **karyogamy**
  the fusion of nuclei of two cells, as part of syngamy.

- **karyotype**
  the appearance of the metaphase chromosomes of an individual or species; comparative size, shape and morphology of the different chromosomes.

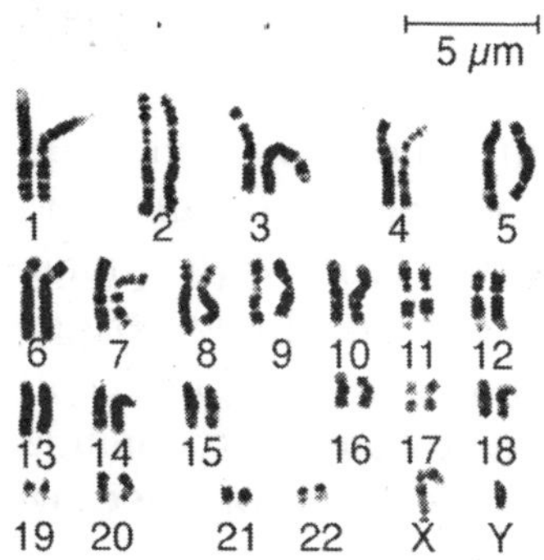

- **karyotype**
  a method of classifying the chromosomes of a cell in relation to number, size and type.

- **key stimulus**
  the stimulus which is adequate to activate an innate releasing mechanism which then sets in motion a species specific action pattern. The term is based on the idea that the releasing cues of a behaviour are analogous to a key which opens a lock.

- **kin recognition**
an animal's ability to differentiate between relatives and non-related conspecific. It is known to occur in high vertebrates, especially mammals and birds, and also sporadically in invertebrates (desert wood lice, honeybees). It may include personal recognition of single individuals; collective recognition of members of a genetically related group by some mechanism such as group odour or hive odour or phenotype matching, in which an animal distinguishes between others on the basis of similarity to itself. See also **phenotype matching.**

- **kin selection**
1. a form of genetic selection whereby alleles differ in their rate of propagation by influencing the survival or reproduction of individuals (kin) who carry the same alleles by common descent.
2. selection for traits that lower an individual's personal fitness, but raise a relative's fitness.
3. a phenomenon of inclusive fitness, used to explain altruistic behaviour between related individuals.

- **kinematic diagram**
a diagram showing the sequencing of behaviour and giving the transition frequencies (or probabilities).

- **kinesis**
non-directional orienting reactions in the presence of a particular sensory stimulus. Animals which suddenly find themselves in an unfavourable environment (e.g. with regard to humidity, temperature, or salt concentration) may change direction by trial and error. By such a method paramecia or woodlice become hedged in an area with favourable environmental conditions.

- **kinesthetic sense or kinesthesia**
sensations of movement and body position an animal's perception of its own movement and posture. Also called idiothetic sense.

- **kinetic energy**
the energy of motion. Moving matter does work by transferring some of its kinetic energy to other matter.

- **kinetochore**
a specialised region on the centromere that links each sister chromatid to the mitotic spindle.

- **kingdom**
a taxonomic category, the second broadest after domain.

- **kinopsis**
individuals of a group are attracted to other group members by visual perception of their movement.

- **kleptogamy**
sneaky or stolen matings.

- **kleptoparasite**
an animal that steals food gathered by another animal.

- **kleptoparasitism**
the stealing of food. Usually refers to interspecific food thievery.

- **klinokinesis**
refers to kinesis orientation behaviour in which the turning rate changes under different conditions.

- **klinotaxis**
a simple type of orientation behaviour in which the animal makes successive comparisons of the intensity of a stimulus on one side of the body and then on the other.

- **krebs cycle**
a chemical cycle involving eight steps that completes the metabolic breakdown of glucose molecules to carbon dioxide; occurs within the mitochondrion; the second major stage in cellular respiration. Also called citric acid cycle or tricarboxylic acid (TCA) cycle.

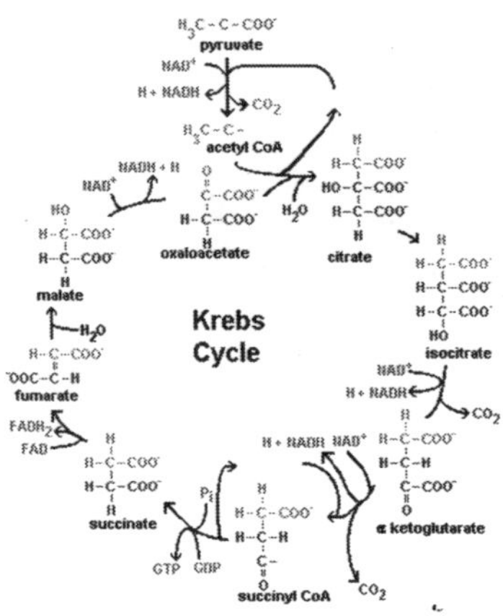

- **k-selection**
selection that favours organisms producing few offspring, each with a relatively high probability of survival; often associated with high competitive ability, long lifespan, slow devel-

opment, late reproduction and poor dispersal ability.

- **kurtosis**
  a deviation from the shape of the normal frequency distribution.
- **kwashiorkor**
  type of malnutrition that occurs in infants and very young children when they are weaned from mother's milk to a starchy diet low in protein.

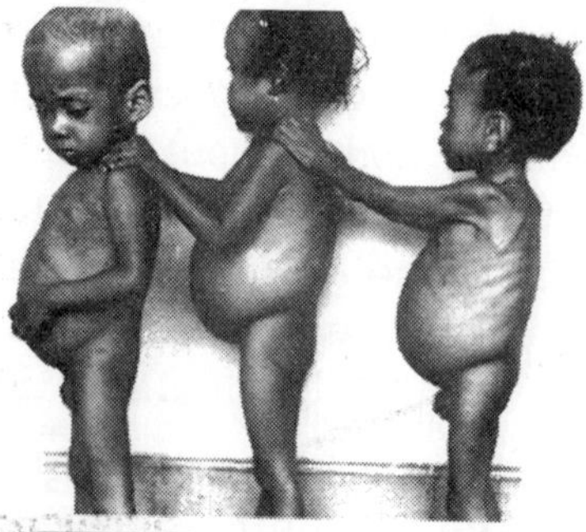

- **labile**
  of a character that changes frequently or readily over evolutionary time.
- **ladder fuels**
  vegetation located below the crown level of forest trees which can carry fire from the forest floor to tree crowns. Ladder fuels may be low-growing tree branches, shrubs or smaller trees.

- **lagging strand**
  a discontinuously synthesised DNA strand that elongates in a direction away from the replication fork.
- **lamarck**
  proposed, in the early 1800s, that evolutionary change may occur via the inheritance of acquired characteristics. This idea, which has since been discredited, holds that the changes in characteristics which occur during an individual's life can be passed on to its offspring.
- **lamarckism**
  the belief that patterns of resemblance among organisms arise through evolutionary modification where animals that resemble each other most closely have close common ancestors. However, the mechanism by which modifications occur is the direct calling forth of adaptive responses in animals which are

then inherited by their offspring; inheritance of acquired characteristics.

- **land use planning**
  the process of organising the use of lands and their resources to best meet people's needs over time, according to the land's capabilities.

- **landing**
  any place where cut timber is assembled for further transport from the timber sale area.

- **landmarks**
  conspicuous natural objects which are useful for optical orientation.

- **landscape**
  a large land area composed of interacting ecosystems that are repeated due to factors such as geology, soils, climate and human impacts. Landscapes are often used for coarse grain analysis.

- **land-use planning**
  process for deciding the best present and future use of each parcel of land in an area.

- **language**
  to qualify as language, a communication system must be culturally transmitted and symbolic; to some it must be capable of encoding abstract ideas.

- **larder hoarding**
  the hoarding of food in a central and mass location. The accumulation of food in one mass location.

- **larva (pl. larvae)**
  a free-living, sexually immature form in some animal life cycles that may differ from the adult in morphology, nutrition and habitat.

- **late forest succession**
  the stage of forest succession in which most of the trees are mature or over mature.

- **late successional plant species**
  mostly trees that can tolerate shade and that form a relatively stable complex forest community.

- **latency of response**
  the measure of time elapsing between the onset of a stimulus and the beginning of the response to it.

- **law of segregation**
  mendel's first law, stating that allele pairs separate during gamete formation, and then randomly re-form pairs during the fusion of gametes at fertilisation.

- **ld cycle**
  a cycle of alternating light and dark periods over a day; LD 16:8 means 16 hours of light alternating with 8 hours dark.

- **leaf area index**
  the area of leaves exposed over a unit area of land surface.

- **learn**
  to exhibit a change in behaviour between two successive exposures to the same environment that cannot be attributed to manipulation of drive operations, alterations in the environment, sensory adaptation, disease, surgical interference, physical trauma, or growth—although the propriety of these exclusions may be questioned.

- **learned helplessness**
  a phenomenon that may occur when avoidance or reduction of some aversive stimulation is impossible or extremely difficult for the subject to avoid. Characterised by passive acceptance of the aversive stimulus.

- **learned preference**
  any preference influenced by a conditioning process.

- **learning**
  an adaptive change in behaviour resulting from experience.

- **learning curve**
  graphic illustration of changes in an animal's responsiveness to the learning process; arranged by consecutive trials.

- **learning set**
  mental representation of a principle common to a series of learning trials such that although circumstances may differ between trials, the solution to each is governed by the common principle. An animal is considered to have acquired a

learning set when it apprehends the principle.

- **lee-boot effect**
  suppression of initiation of estrous behaviour by overcrowding of animals.

- **lek**
  communal mating area in which individuals hold small territories solely for courtship and copulation.

- **leptokurtic**
  a symmetrical frequency distribution which differs from a normal distribution by deficiency in the shoulders compared to the tails and the top.

- **lesion**
  in ethology and physiology, deliberate destruction of neural tissue for experimental purposes, as by surgical severing of a nerve tract in the brain.

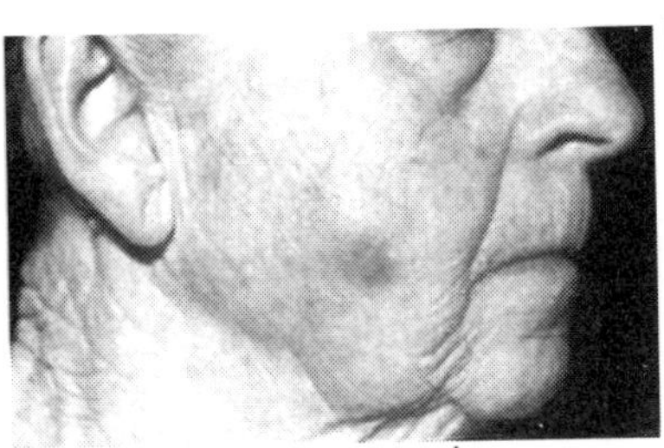

- **lestobiosis**
  the relation in which colonies of small species of social insects nest in the walls of the nests of a larger species and enter the chambers of the larger species to prey on brood or to rob the food stores.

- **lethal**
  a lethal mutation or aberration causes the death of the individuals possessing it; those that can be studied genetically are lethal only when homozygous.

- **leukocyte**
  a white blood cell; typically functions in immunity, such as phagocytosis or antibody production.

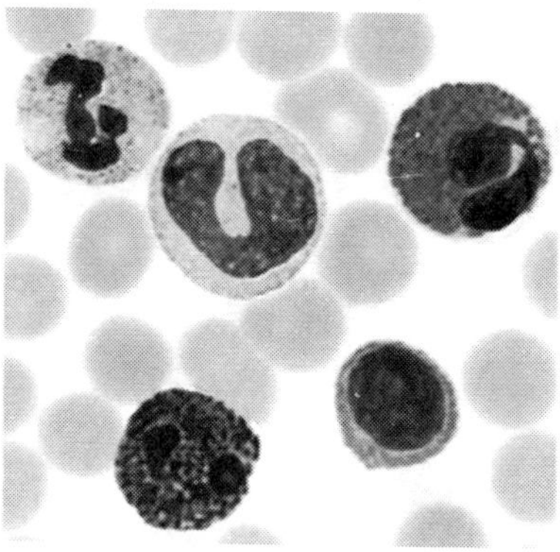

- **levels of organisation**
  a basic concept in biology is that organisation is based on a hierarchy of structural levels, with each level building on the levels below it.

- **lichen**

  1. an organism formed by the symbiotic association between a fungus and a photosynthetic alga.
  2. a *symbiotic* relationship between a fungus and a moss. The moss does most of the work, producing sugars for the lichen's collective metabolic pathways. Lichen are generally low-growing, vary in colour from bright orange or yellow to grey or black, and are often found growing on rocks and tree bark.

- **licking**

  an action whereby the tongue is slid over the surface of an object. Licking may have a nutritional, gustatory, prophylactic or social function; may be self directed or social.

- **life**

  a table of data summarising mortality in a population.

- **life cycle**

  the sequence of stages through which an organism passes in development from a zygote to the production of progeny zygotes.

- **life form**

  characteristic structure of a plant or animal.

- **life history**

  an organism's lifetime pattern of growth, differentiation, storage and reproduction.

- **life table**

  a summary of the age- or stage-related survivorship of individuals in a population.

- **life-history strategy**

  a selected set of adaptations to local environments, involving such quantitative aspects of life history as fecundity, the timing of maturation and the frequency of breeding.

- **ligament**

  a type of fibrous connective tissue that joins bones together at joints.

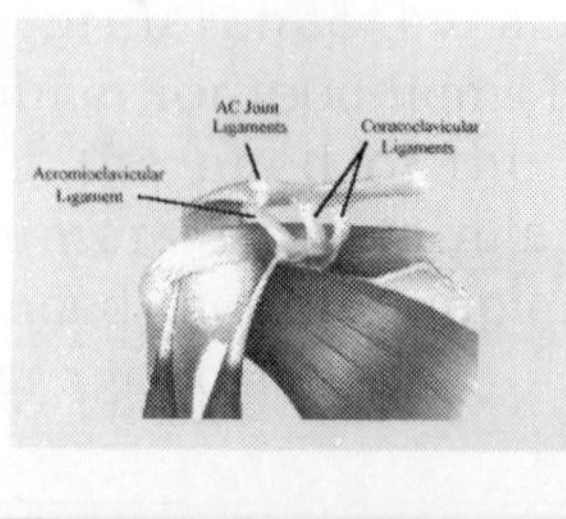

- **ligand**
  a ligand is a molecule that binds specifically to a receptor site of another molecule. A ligase is an enzyme which catalyses such a reaction.

- **light adaptation**
  process in which the sensitivity of the photoreceptor response decreases when exposed to light, so that each additional photon generates a smaller response.

- **light compass reaction**
  locomotion at a temporarily fixed angle to light rays.

- **lignin**
  complex organic material deposited within the cell walls of plants associated with cellulose, especially in wood and fibres.

- **lignophagia**
  eating wood.

- **limb flicking**
  a behaviour of cats when they get their paws wet. Each wet limb is quickly shaken in turn.

- **limiting similarity**
  the level of similarity between two competing species which cannot be exceeded if the species are to coexist.

- **lineage**
  line of descent.

- **linkage**
  the occurrence of genes on the same chromosome. The closer they are on the chromosome, the more tightly they are linked, that is, the less likely they are to segregate independently during meiosis.

- **linkage equilibrium and linkage disequilibrium**
  if two alleles at two or more loci are associated more or less frequently than predicted by their individual frequencies, they are in linkage disequilibrium; if not, they are in linkage equilibrium.

- **linked genes**
  genes that are located on the same chromosomes.

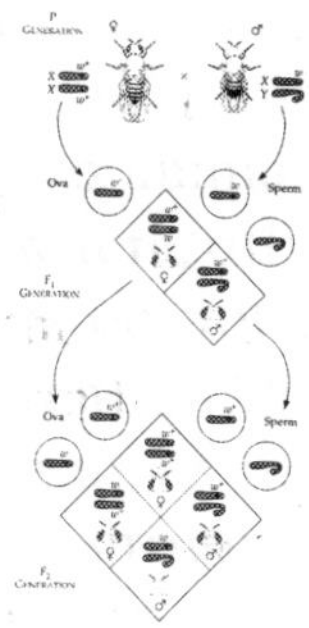

- **lipase**
  enzyme(s) that split true fats into alcohols and acids.

- **lipid**
  one of a family of compounds, including fats, phospholipids and steroids, that are insoluble in water.

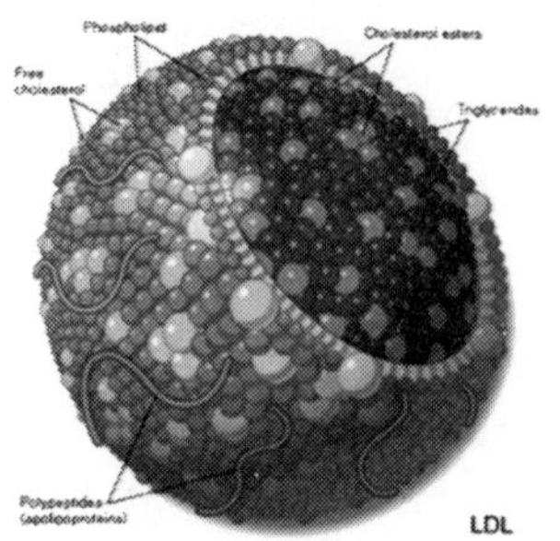

- **liposome**
  liposomes are vesicles (spherules) in which the lipid molecules are spontaneously arranged into bilayers with hydrophilic groups exposed to water molecules both outside the vesicle and in the core.

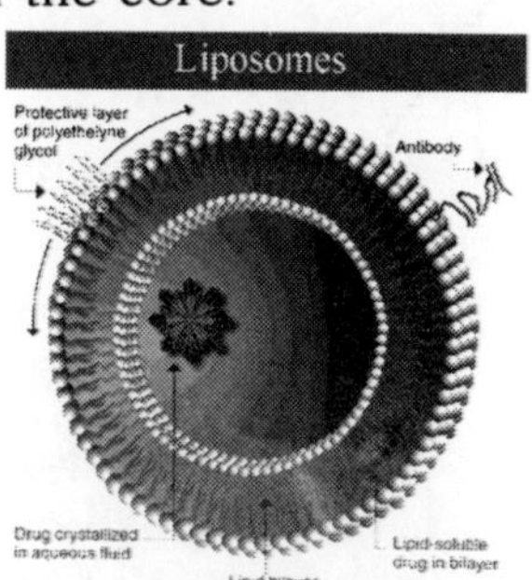

- **lithosphere**
  solidified rock surface or outer shell of the earth.

- **litter (forest litter)**
  the freshly fallen or only slightly decomposed plant material on the forest floor. This layer includes foliage, bark fragments, twigs, flowers and fruit.

- **littoral forest**
  forests situated along the coastline, such as mangroves and eucalypts located immediately adjacent to the beach.

- **littoral zone**
  the zone at the edge of a lake or ocean which is periodically exposed to the air and periodically immersed.

- **local enhancement**
  an animal's directing of attention toward some object or place in its environment as a consequence of observing how others act with regard to that object or place. Thus many animals find their way to a source of food more quickly if they see others looking for that place. Like social facilitation, local enhancement is to be distinguished from imitation.

Each animal finds its own way by trial and error, even though the process is facilitated by observing other animals.

- **local stability**

  the tendency of a community to return to its original state when subjected to a small perturbation.

- **locomotion**

  voluntary movements which displace the whole body of an organism.

- **locus**

  a particular place along the length of a certain chromosome where a specified allele is located.

- **logging residue**

  the residue left on the ground after timber cutting. It includes unutilised logs, uprooted stumps, broken branches, bark and leaves. Certain amounts of slash provide important ecosystem roles, such as soil protection, nutrient cycling and wildlife habitat.

- **logistic equation**

  1. a much-used equation for the growth of a single-species population without discrete generations under the influence of competition, which is simple and captures the essence of actual examples but very few of the details of actual examples.

  2. a specific equation describing the idealised growth of a population subject to a density-dependent limiting factor.

- **logistic growth**

  a pattern of population growth in which the growth rate decreases with an increasing number of individuals (S-shaped curve).

- **logistic population growth**

  a model describing population growth that levels off as population size approaches carrying capacity.

- **long call**

  in gulls a demonstrative behaviour with accompanying vocalisation. In the herring gull the sequence occurs in three parts and consists of stretching the head first forward under the breast then backward while uttering a few high notes, then the head is jerked upward, a series of loud sounds are emitted followed by thrusting the neck and wide open beak forward.

- **long-day plant**
  a plant that flowers, usually in late spring or early summer, only when the light period is longer than a critical length.

- **longevity**
  length of life.

- **loop of Henley**
  the long hairpin turn, with a descending and ascending limb, of the renal tubule in the vertebrate kidney; functions in water and salt reabsorption.

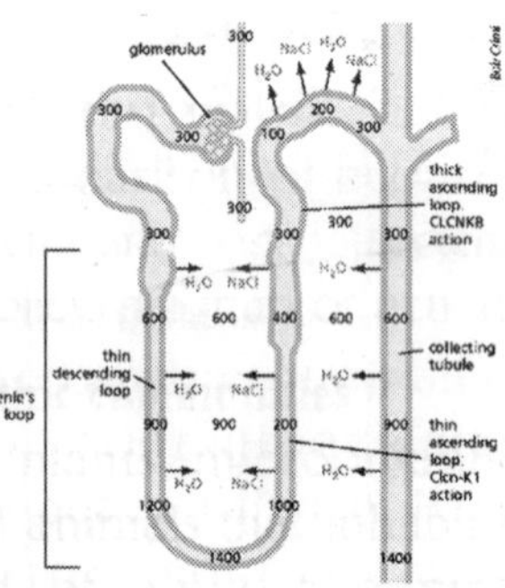

- **lordosis**
  the posture adopted by many female mammals before and during copulation. Can be elicited by tactile stimulation.

- **lunar periodicity**
  in many marine animals living in the tidal zone, the reproductive cycles are limited to the periods of spring tide and may coincide precisely with the moon phases.

- **lying-out period**
  in ungulates the young will actively leave the mother after nursing and hide, concealed until called by the mother.

- **lymph**
  the colourless fluid, derived from interstitial fluid, in the lymphatic system of vertebrate animals.

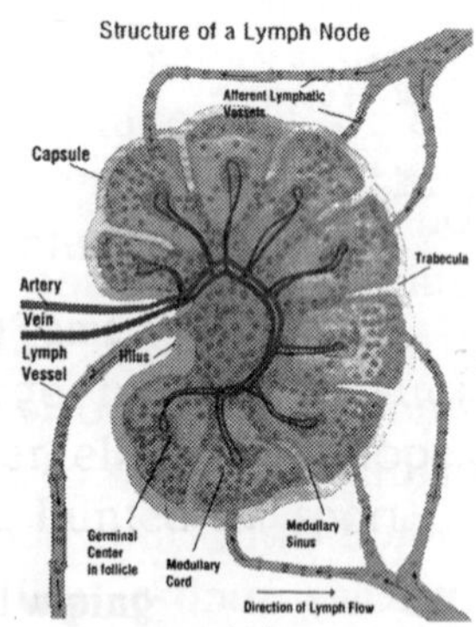

- **lymph nodes**
  the lymphatic system is a complex network for the distribution of lymph fluid (which is similar to blood

plasma -blood without red cells). Lymph is collected by drainage from the tissues throughout the body, flows in the lymphatic vessels through the lymph nodes, and is eventually added to the venous blood circulation. Lymph consists of a clear liquid portion, varying numbers of white blood cells (chiefly lymphocytes) and a few red blood cells. The lymph nodes are small bodies located throug-hout the lymph system and varying in diameter from 0.1 to 2.5 centimetres.

- **lymphatic system**

a system of vessels and lymph nodes separate from the circulatory system that returns fluid and protein to the blood.

- **lymphocyte**

lymphocytes are a type of leukocyte (white blood cell) responsible for the immune response. There are two classes of lymphocytes:

1. the B-cells, when presented with a foreign chemical entity (antigen), change into antibody producing plasma cells;

2. the T-cells interact directly with foreign invaders such as bacteria and viruses. The T-cells express various surface marker macromolecules. For example, CD4+ is the notation for a specific expressed T-cell surface marker that can be identified by assay.

3. a white blood cell. The lymphocytes that complete their development in the bone marrow are called B cells, whereas the lymphocytes that mature in the thymus are called T cells. The former are responsible for humoral immunity; the latter are responsible for cell-mediated immunity.

- **lysosome**

a membrane-bounded organelle found in eukaryotic cells (other than plants). Lysosomes contain a mixture of enzymes that can digest most of the macromolecules found in the rest of the cell.

Lysosome Structure

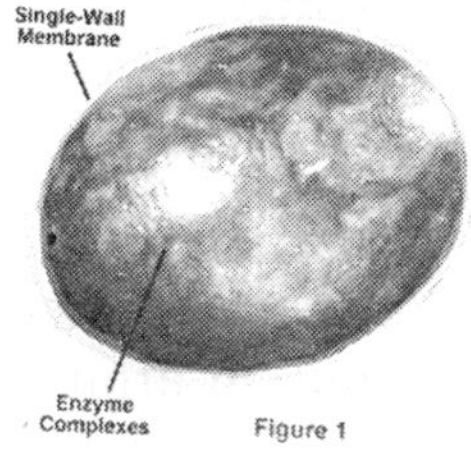

Figure 1

- **m**

1. the migration rate and population genetics; the fraction of a population that consists of individuals that immigrated to it from other populations.
2. five thousand board feet of timber can be expressed as 5M board feet.

- **m phase**

the mitotic phase of the cell cycle, which includes mitosis and cytokinesis.

- **macro climate**

the general, large scale climate of a large area, as distinguished from the smaller scale micro climates within it.

- **macroevolution**

a vague term for the evolution of great phenotypic changes, usually great enough to allocate the changed lineage and its descendants to a distinct genus or higher taxon, evolution above the species level. The study of evolutionary events and processes that require long times for their occurrence or operation.

- **macrofauna**

large animals in a community within an arbitrary size range, e.g. between 2 mm and 20 mm body width, in soil invertebrates.

- **macroinvertebrate**

an invertebrate with a body length greater than 2 mm.

- **macromolecule**

a giant molecule of living matter formed by the joining of smaller molecules, usually by condensation synthesis. Polysaccharides, proteins and nucleic acids are macromolecules.

- **macromutationism**

a theory that new taxa evolve essentially instantaneously via a major genetic mutation that establishes reproductive isolation and new adaptations all at once.

- **macroparasite**

a parasite which grows in its host but multiplies by producing infective stages that are released from the host to infect new hosts. They are often intercellular or live in body cavities, rather than within the host cells.

- **macrophage**

an amoeboid cell that moves through tissue fibres,

engulfing bacteria and dead cells by phagocytosis.

- **magma**
  molten or partly molten rock material.

- **magnetic field orientation**
  animal orientation using the earth's magnetic field.

- **magnitude of response**
  term applied to any one of several descriptive measures of a response, including amplitude or duration, latency or velocity of response, or of a temporally restricted set of responses (such as frequency, relative frequency, percentage of occurrence), or rate of response, which state quantitatively the likelihood that a response will occur under stated conditions during that brief span of time.

- **maintenance activities**
  behaviour patterns and movements that serve to enhance the animal's comfort and body care including cleaning movements, self-shaking movements, self rubbing, bathing movements, stretching, scratching and yawning and defecation.

- **Major Histocompatibility Complex or MHC**
  a large set of cell surface antigens encoded by a family of genes. Foreign MHC markers trigger T-cell responses that may lead to rejection of transplanted tissues and organs.

- **maladaptive behaviour**
  any behaviour that causes a reduction in fitness.

- **male care**
  parental care performed by males.

- **male choice**
  non-random mating by males.

- **male parasitism**
  the permanent attachment of a male to a female.

- **malignant tumour**
  a cancerous growth; an abnormal growth whose cells multiply excessively, have altered surfaces and may have unusual numbers of chromosomes and/or aberrant metabolic processes.

- **malphighian tubule**
  a unique excretory organ of insects that empties into the digestive tract, removes nitrogenous wastes from the

blood and functions in osmoregulation.

- **mammal**
  a taxon of animals characterised by the presence of hair and mammary glands (females).

- **mammary**
  a milk-producing gland found in female mammals.

- **mean**
  usually the arithmetic mean or average; the sum of n values, divided by n.

- **mean annual increment of growth**
  the total increase in size or volume of individual trees. Or, it can refer to the increase in size and volume of a stand of trees at a particular age, divided by that age in years.

- **mean intensity of infection**
  the mean number of parasites per host in a population (including those hosts that are not infected.)

- **measurement of response**
  the assignment of numbers to responses or to response-instances according to the rules of measurement. Measures include amplitude of response, intensity of response, latency of responses, probability of response, rate of response.

- **mechanical isolation**
  reproductive isolation owing to the incompatibility of male and female genitalia.

- **mechanical senses**
  mechanoreceptors are receptors that are sensitive to mechanical deformation.

- **mechanoreceptor**
  1. sensory cell that reacts to mechanical deformation.
  2. sensory receptor for mechanical (movement) energy.
  3. a sensory receptor that detects physical deformations in the body environment associated with pressure, touch, stretch, motion and sound.

- **medulla oblongata**
  the lowest part of the vertebrate brain; a swelling of the hindbrain dorsal to the anterior spinal cord that controls autonomic, homeostatic functions, including breathing, heart and blood vessel activity, swallowing, digestion and vomiting.

- **medusa**
  the floating, flattened, mouth-down version of the cnidarian body plan. The alternate form is the polyp.

- **megapascal**
  a unit of pressure equivalent to 10 atmospheres of pressure.

- **megaphyte**
  plants with normally unbranched stems or trunks and bearing a crown of very large leaves. The inflorescence is commonly also massive.

- **meiosis**
  a type of nuclear division in which the number of chromosomes is reduced to half.

- **meiotic drive**
  mechanisms by which some genes end up preferentially in the daughter cells at meiosis.

- **melatonin**
  a secretion of the pineal gland which seems to play a role in the sleep/wake cycle. Function of the pineal gland appears to be influenced by the hypothalamus.

- **meliforous**
  honey producing.

- **membrane potential**
  the charge difference between the cytoplasm and extra cellular fluid in all cells, due to the differential distribution of ions. Membrane potential affects the activity of excitable cells and the transmembrane movement of all charged substances.

- **membrane resistance**
  property of the cell membrane that impedes the flow of current.

- **meme**
  the unit of information transfer be it idea, belief or custom of cultural evolution (analogous to gene in genetical evolution.)

- **memory**
  the capability of the CNS to store information for subsequent retrieval.

- **mendelian inheritance**
  a theory of inheritance that states that the hereditary traits are carried in cells in the form of particles (genes).

- **menstrual cycle**
  1. refers to the repeated pattern of periodic bleeding from the uterus and vagina which occurs in sexually

mature women, in all species of apes and in some kinds of monkeys.
2. a type of reproductive cycle in higher female primates, in which the non-pregnant endometrium is shed as a bloody discharge through the cervix into the vagina.

- **menstruation**
vaginal bleeding resulting from shedding of the uterine lining during a menstrual cycle.

- **meristem**
a part of a plant in which cell division is concentrated (especially the shoot apices, lateral buds and a zone near the tips of roots and rootlets.)

- **meristic trait**
a discretely varying, countable trait; e.g., number of digits.

- **meroblastic cleavage**
a type of cleavage in which there is incomplete division of yolk-rich egg, characteristic of avian development.

- **mesenteries**
memranes that suspend many of the organs of vertebrates inside fluid-filled body cavities.

- **mesoderm**
the middle primary germ layer of an early embryo that develops into the notochord, the lining of the coelom, muscles, skeleton, gonads, kidneys and most of the circulatory system.

- **mesofauna**
animals in the size-range 100 mm to 2 mm body length.

- **mesophyll**
1. the internal non-vascular tissue of a leaf. In green plants it is in these cells that most chloroplasts are found and in which most photosynthesis occurs.
2. the ground tissue of a leaf, sandwiched between the upper and lower epidermis and specialised for photosynthesis.

- **mesophyll cell**
a cell in a leaf's bundle sheath.

- **mesosome**
a localised infolding of the plasma membrane of a bacterium.

- **messenger**
(RNA) a type of RNA synthesized from DNA in the genetic material that attaches to ribosomes in the cytoplasm and specifies the primary structure of a protein.

- **metabolism**
1. the sum of all chemical reactions occurring within a cell or an organism.
2. the sum total of the chemical and physical changes constantly taking place in living substances.

- **metabotropic receptor**
membrane receptor that binds a neurotransmitter and thereby begins a sequence of biochemical reactions in second messenger systems that ultimately cause a slow change in the electrical properties or physiological state of the cell.

- **metacommunication**
a form of communication that conveys information about the interpretation of subsequent communications or signals.

- **micro climate**
the climate of a small site. It may differ from the climate at large of the area due to aspect, tree cover (or the absence of tree cover), or exposure to winds.

- **microbes**
microorganisms: any microscopic organism, including bacteria, viruses, unicellular algae and protosoans and microscopic fungi such as yeasts.

- **microbivores**
animals that feed on microorganisms.

- **microbody**
a small organelle, bounded by a single membrane and possessing a granular interior. Peroxisomes and glyoxysomes are types of microbodies.

- **microclimate**
the climate within a very small area or in a particular, often tightly defined, habitat.

- **microevolution**
a change in the gene pool of a population over a succession of generations.

- **microfauna**
the smallest arbitrary size categorisation of animals in a community.

• **microfilament**
minute fibrous structure generally composed of action found in the cytoplasm of eukaryotic cells. They play a role in motion within cells.

• **microflora**
bacteria, fungi and microscopic algae.

• **microparasite**
a parasite which multiplies directly within its host, usually within host cells.

• **microscope**
an instrument which magnifies images either by using lenses in an optical system to bend light (light microscope) or electromagnets to direct the movement of electrons (electron microscope.)

• **microsite**
the small subset of environments within a habitat that provide the specialised resources and conditions required for a phase in the life of an organism, e.g. the cracks or crevices which provide conditions suitable for germination of the seed of a particular species.

• **microtubule**
a minute tubular structure found in centrioles, spindle apparati, cilia, flagella and other places in the cytoplasm of eukaryotic cells. Microtubules play a role in movement and maintenance of shape.

• **microvillus**
collectively, fine, finger like projections of the epithelial cells in the lumen of the small intestine that increase its surface area.

• **middle lamella**
a thin layer of adhesive extra cellular material, primarily pectins, found between the primary walls of adjacent young plant cells.

• **middle ground**
a term used the management of visual resources, or scenery. It refers to the visible terrain beyond the foreground where individual trees are still visible but do not stand out distinctly from the stand.

• **migration**
the movement of animals in response to seasonal changes or changes in the food supply. Examples of animals that

migrate include ruby-throated hummingbirds, salmon, monarch butterflies, buffalo and elephants.

- **migratory restlessness**

activity of migratory birds held in captivity; a good measure of readiness to migrate.

- **Miller and Urey**

conducted pioneering work which documented a possible mechanism which allowed for the production of monomers (including amino and fatty acids) under conditions existing at the earth's surface before life forms existed.

- **mimetic behaviour**

an animal is placed in the environment of another animal of the same species: if the second animal then makes some species-specific response, and thereafter the first animal makes the same response under conditions where it can be demonstrated that the response of the second animal is the stimulus for that of the first and that no opportunity for discrimination and differentiation training has been given to either animal, then the behaviour is termed mimetic.

- **mimicry**

1. the resemblance of an organism (the mimic) either to another organism or to a non-living object (the model), presumably conferring a benefit on the mimic in natural selection.
2. imitation, deception, disguise and protective adaptation to make an individual appear to be something it is not. The superficial resemblance of one organism by another, presumably affecting the actions of predators of prey.
3. a phenomenon in which one species benefits by a superficial resemblance to an unrelated species. A predator or species of prey may gain a significant advantage through mimicry.

- **mineral soil**

soil that consists mainly of inorganic material, such as weathered rock, rather than organic matter.

- **miocene**

1. a geological era lasting from approximately 25 to 5 million years ago.
2. a middle epoch of the Tertiary period, beginning

about 26 million years ago and ending about 11 million years ago.

- **MIS (Management Indicator Species)**
  a wildlife species whose population will indicate the health of the ecosystem in which it lives and, consequently, the effects of forest management activities to that ecosystem. MIS species are selected by land management agencies.

- **mitigation**
  actions taken to avoid, minimise or rectify the impact of a land management practice.

- **mitochondria**
  cytoplasmic organelles composed of two membranes, the innermost of which is convoluted; they are the site of many reactions of cellular respiration.

- **mitochondrial matrix**
  the compartment of the mitochondrion enclosed by the inner membrane and containing enzymes and substrates for the Krebs cycle.

- **mitochondrion**
  an organelle that occurs in eukaryotic cells and contains the enzymes of the citric acid cycle, the respiratory chain and oxidative phosphorylation. A mitochondrion is bounded by a double membrane.

- **mitosis**
  the orderly processes through which the nucleus of the cell undergoes division in such a fashion that the daughter nuclei receive the same number of chromosomes as in the parent cell.

- **mixed species group**
  a group composed of more than one species.

- **mixed stand**
  a stand consisting of two or more tree species.

- **mixed strategy**
  more than one strategy is maintained in the population.

- **mobbing**
  aversive response by song birds to birds of prey in which they swarm around and pursue their enemy while giving typical calls and performing sham attacks.

- **modal action patterns**
recognising that fixed action patterns always have some variability. Barlow argued that they should be referred to as modal rather than fixed.

- **model**
two totally different meanings:
1. a simulation of some natural object, such as a mannequin or decoy. These are stimulus objects used in behavioural experiments to determine the key stimuli, releasers, supernormal releasers or generally the effects of a stimulus object on the behaviour of an organism.
2. an hypothesis or paradigm constructed for the purpose of testing a scientific idea.

- **modern synthesis**
a comprehensive theory of evolution emphasizing natural selection, gradualism, and populations as the fundamental units of evolutionary change; also called Neo-Darwinism.

- **modifier genes**
genes whose major obvious phenotypic effect is to modify the phenotypic expression of other genes.

- **mouth play**
the young of many species will play bite, nip or mouth body parts of others as they play.

- **mucosa**
refers to the mucous tissue lining various tubular structures in the body.

- **mucous membrane**
a type of epithelium that secretes a lubricant, lining the digestive tract and air tubes.

- **mullerian mimicry**
1. a mutual mimicry by two unpalatable species.
2. mimicry in which a series of dangerous or distasteful species resemble one another.

- **multigene family**
a collection of genes with similar or identical sequences, presumably of common origin.

- **multiparous**
females that have undergone more than one cycle of pregnancy and parturition.

- **multiple fruit**
a fruit, such as a pineapple, that develops from an inflorescence, a group of flowers clustered tightly together.

- **multiple resistance**
resistance of an organism to a number of pesticides requiring different mechanisms to counteract their effects.

- **multiple use management**
the management of all the various renewable surface resources of National Forest lands for a variety of purposes such as recreation, range, timber, wildlife and fish habitat and watershed.

- **multipolar neuron**
neuron with a single long axon and many dendrites extending form the cell body.

- **mushroom body**
brain region of insects containing Kenyon cells; may be implicated in more complex types of behaviour such as learning, social behaviour and spatial memory.

- **mutagen**
a chemical or physical agent that interacts with DNA and causes a mutation.

- **mutagenesis**
the creation of mutations.

- **mutagenic agent**
an agent leading to an increase in mutation rate above the spontaneous level at a locus.

- **mutation**
a change in a gene; sometimes applied also to chromosomal changes. An error in replication of a nucleotide sequence, or any other alteration of the genome that is not manifested as reciprocal recombination.

- **mutation pressure**
changes in allelic frequencies in a population due to mutation.

- **mutation rate**
the frequency of mutations per generation at a given locus.

- **mutual antagonism**
describes two species with reciprocal negative effects on each other (either interspecific competition or mutual predation).

- **mutual interference**
interference amongst predators leading to a reduction in the consumption

rate of individual predators which increases with predator density.

- **mutualism**

1. an interaction between two organisms (conspecifics or heterospecific) which benefits (increases the reproductive success of) both interactants. A relationship between two or more individuals or species in which both parties benefit. Normally benefits are measured by their net effects on direct fitness.

2. a symbiotic relationship in which both the host and the symbiont benefit.

- **mycelium**

the densely branched network of hyphae in a fungus.

- **mycorrhizae**

mutualistic associations of plant roots and fungi.

- **myelin**

a layer of cells that cover nerve fibres and increase conduction velocity.

- **myelin sheath**

an insulating coat of cell membrane from schwann cells that is interrupted by nodes of Ranvier where saltatory conduction occurs.

- **myofibrils**

fibrils arranged in longitudinal bundles in muscle cells (fibres); composed of thin filaments of actin and a regulatory protein and thick filaments of myosin.

- **myoglobin**

an oxygen-storing, pigmented protein in muscle cells.

- **myosin**

a type of protein filament that interacts with actin filaments to cause cell movement, such as contraction in muscle cells.

- **myrmecophagous**

eating ants.

- **myrmecophilous**

living in association with ants.

- **n**

sample size; the number of organisms in the population, population size.

- **nanometer**

a unit of measure (length). 1 nm is equal to 1×10^{-9} m, or 1/1,000,000 mm.

- **nape bite**

1. a bite directed at the dorsal neck; it can be an inhib-

ited bite that does not penetrate the skin, such as that associated with tomcats, dogs and stallions during mating.

2. directional killing bite of a cat on its prey. The bite is aimed at a constriction between two head and shoulders and results in severing the spinal cord.

- **natal**

refers to birth, as in natal sites or birthplace.

- **native species**

a species that occurs naturally in an area or habitat. Also called indigenous species.

- **natural barrier**

a natural feature, such as a dense stand of trees or downfall, that will restrict animal travel.

- **natural resource**

1. a feature of the natural environment that is of value in serving human needs.

2. nutrients and minerals in the soil and deeper layers of the earth's crust; water; wild and domestic plants and animals; air and other resources produced by the earth's natural processes.

- **natural selection**

1. process by which environmental effects lead to varying degrees of reproductive success among individuals of a population of organisms with different hereditary characters, or traits. The characters that inhibit reproductive success decrease in frequency from generation to generation.

2. the process by which genetic traits are passed on to each successive generation. Over time, natural selection helps species become better adapted to their environment. Also known as 'survival of the fittest,' natural selection is the driving force behind the process of evolution.

3. the differential reproduction of genotypes resulting from effects of environmental factors on individuals exhibiting heritable variability. A mechanism of evolution proposed by Darwin and Wallace.

4. one of several gradual mechanisms through which evolution occurs. Process by which a particular beneficial gene (or set of genes)

is reproduced more than other genes in succeeding generations due to selective pressures in the environment that favour the beneficial gene. The result of natural selection is a population that contains a greater proportion of organisms better adapted to certain environmental conditions.

- **naturalistic fallacy**
  the erroneous belief that what is found in nature is right or morally justified.

- **navigation**
  the ability to orient toward a goal (e.g. home) regardless of its direction by means other than recognition of landmarks.

- **nebula**
  interstellar cloud of gas and dust.

- **neck-bite**
  1. in a variety of mammals and birds the male bites the female's neck during copulation.
  2. similar movements occur in many predators when they attack their prey.
  3. this type of neck bite also resembles the grip used by certain mammals and rodents when carrying their young.

- **necromass**
  the weight of dead organisms, usually expressed per unit of land or volume of water. The term is sometimes used to include the dead parts of living organisms, e.g. the bark and heartwood of trees, the hair and claws of animals.

- **necroparasite**
  a parasite that kills its host (or a part of it) and continues growth on the dead resource.

- **necrophagic**
  feeding on dead organisms.

- **necrophoresis or necrophoric behaviour**
  transport of dead members of the colony away from the nest in social insects.

- **nectar guide**
  the lines on a flower that lead to the location where an insect may obtain nectar from that flower.

- **need**
  a state of physiological disequilibrium and of departure from homeostatic balance.

• **needham**
concluded from his experiments that spontaneous generation can occur. However, his experimental protocol was flawed, allowing contaminant micro organisms to proliferate in the experimental system.

• **negative feedback**
1. a dynamic relation whereby the product of a process inhibits the process that produces it, usually enhancing stability.
2. a primary mechanism of homeostasis, whereby a change in a physiological variable that is being monitored triggers a response that counteracts the initial fluctuation.

• **negative feedback loop**
situation in which a change in a certain direction provides information that causes a system to change less in that direction. This is a common regulatory mechanism and is widely used in animals to control hormone levels in the blood. For example, the hormones that control ovulation in humans are on a negative feedback loop.

• **neo-darwinism**
the addition of modern genetics to Darwin's theory.

• **neo-lamarckism**
a popular late 19th century alternative to Darwinism, postulating that adaptations arise as characters acquired by active organic responses to environments and are passed on to offspring by heredity.

• **neonatal**
newborn.

• **neontology**
the study of recent organisms.

• **neophobia**
a fear by animals of unfamiliar objects or situations.

• **neoteny**
slowing down of development over evolutionary time where animals become sexually mature before it is physically grown.

• **nephron**
the tubular excretory unit of the vertebrate kidney.

- **nepotism**
  the preferential donating of care of resources to close relatives.

- **neutron**
  an uncharged subatomic particle of about the same size and mass as a proton.

- **niche**
  1. the limits, for all important environmental features, within which individuals of a species can survive, grow and reproduce.
  2. the range of each environmental variable, such as temperature, humidity and food items, within which a species can exist and reproduce. The preferred niche is the one in which the species performs best, and the realized niche is the one in which it actually comes to live in a particular environment.

- **niche differentiation**
  the tendency for coexisting species to differ in their niche requirements.

- **niche packing**
  the tendency for coexisting species between them to fill the available 'space' along important niche dimensions.

- **nidicolous**
  altricial bird chicks that remain in the nest for extended periods of time and are dependent on parental care.

- **nidification**
  nest building.

- **nidifugous**
  baby birds that are generally well developed when hatched and rely on their parents primarily to lead them to food.

- **nitrification**
  the conversion of nitrites to nitrates, usually by micro organisms. The term is commonly used to describe the process of conversion of ammonium ions via nitrites to nitrates.

- **nitrogen cycle**
  cyclic movement of nitrogen in different chemical forms from the environment to organisms and then back to the environment.

- **nitrogen fixation**
  1. the conversion of gaseous nitrogen (N_2) into more complex molecules. The process is used industrially to produce nitrogen

fertilisers. Biological nitrogen fixation is accomplished by both free-living and symbiotic micro organisms (prokaryotes). The process is more properly called 'dinitrogen fixation'.
2. the assimilation of atmospheric nitrogen by certain prokaryotes into nitrogenous compounds that can be directly used by plants.
3. the process of chemically converting nitrogen gas (N_2) from the air into compounds, such as nitrates (NO_3), nitrites (NO_2), or ammonia (NH_3), that can be used by plants in building amino acids and other nitrogen-containing organic molecules.

- **no action alternative**
the most likely condition expected to exist in the future if management practices continue unchanged.

- **nod swimming**
an epigamic behaviour found in several dabbling ducks originally referred to as 'coquette' swimming.

- **node**
the place on a stem where one or more leaves arise.

- **nodes of ranvier**
the small gaps in the myelin sheath between successive glial cells along the axon of a neuron; also, the site of a high concentration of voltage-sensitive ion channels.

- **nodule**
an enlarged root section where nitrogen fixation takes place in legumes such as in peas or beans; composed of plant cells containing bacteroids with symbiotic bacteria.

- **nomenclature**
the method of assigning names in the classification of organisms.

- **nominal data**
refers to categorising or grouping observations into definable categories when there is no underlying scale, e.g. number of responses to red, green and yellow stimuli. If the stimuli were light, medium and dark then there would be an underlying scale even if it was not known to the observer, i.e. the stimuli could be ordered. These would be ordinal data.

- **nonbiodegradable**
  not able to be consumed and/or broken down by biological organisms. Non-biodegradable substances include plastics, aluminum and many chemicals used in industry and agriculture. Particularly dangerous are nonbiodegradable chemicals that are also toxic and tend to accumulate in organisms.

- **noncommercial vegetative treatment**
  the removal of trees for reasons other than timber production.

- **noncompetitive inhibitor**
  a substance that reduces the activity of an enzyme by binding to a location remote from the active site, changing its conformation so that it no longer binds to the substrate.

- **nonconsumptive use**
  the use of a resource that does not reduce the supply. For instance, bird watching is a non-consumptive use of wildlife. Boating and fishing are non-consumptive uses of water.

- **noncyclic electron flow**
  a route of electron flow during the light reactions of photosynthesis that involves both photosystems and produces ATP, NADPH and oxygen; the net electron flow is from water to NADP+.

- **noncyclic photophosphorylation**
  the production of ATP by noncyclic electron flow.

- **nondescendant kin**
  relatives other than offspring.

- **nondisjunction**
  1. failure of the chromosomes to separate at the first meiotic anaphase.
  2. an accident of meiosis or mitosis, in which both members of a pair of homologous chromosomes or both sister chromatids fail to separate normally.

- **non-equilibrium theory**
  in community ecology, concerned with the transient behaviour of a system away from any equilibrium point, it specifically focuses attention on time and variation.

- **nongame**
  wildlife species that are not hunted for sport.

- **nonparametric statistics**
  statistics that do not depend on any particular underlying distribution.

- **nonpoint source pollution**
  pollution whose source is not specific in location. The sources of the discharge are dispersed, not well defined, or constant. Rain storms and snowmelt often make this type of pollution worse. Examples include sediments from logging activities and runoff from agricultural chemicals.

- **nonpolar**
  electrically symmetrical. For example, in many molecules with covalent bonds, the electrons are shared equally; the poles are electrically neutral.

- **non-renewable resource**
  a resource whose total quantity does not increase measurably over time, so that each use of the resource diminishes the supply.

- **non-renewable resource**
  resource that exists in a fixed amount (stock) in various places in the earth's crust and has the potential for renewal only by geological, physical and chemical processes taking place over hundreds of millions to billions of years. Examples are copper, aluminium, coal and oil. We classify these resources as exhaustible because we are extracting and using them at a much faster rate than they were formed.

- **nonsense mutation**
  a mutation that changes an amino acid codon to one of the three stop codons, resulting in a shorter and usually nonfunctional protein.

- **non-spiking neuron**
  neurons function without producing action potentials. The cell is incapable of generating action potentials due to a lack of voltage sensitive ion channels.

- **norm of reaction**
  1. the set of phenotypic expressions of a genotype under different environmental conditions.

2. the range of phenotypic possibilities for a single genotype, as influenced by the environment.

- **notice of intent**
  a notice in the federal register of intent to prepare an environmental impact statement on a proposed action.

- **notochord**
  a longitudinal, flexible rod formed from dorsal mesoderm and located between the gut and the nerve cord in all chordate embryos.

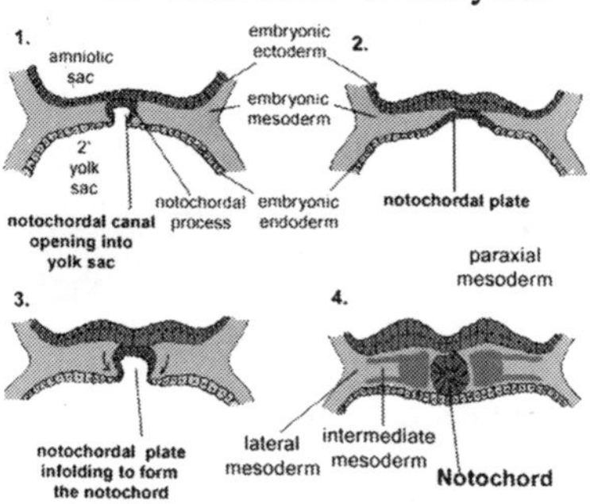

- **novelty**
  an unfamiliar object or situation.

- **'n'-shaped curve**
  more correctly, 'unimodal': a curve with a single maximum.

- **nuclear family**
  a family consisting of the mother, father and their offspring but not including grandparents, grand offspring or other kin.

- **nucleic acid**
  1. a polymer composed of nucleotides that are joined by covalent bonds (phosphodiester linkages) between the phosphate of one nucleotide and the sugar of the next nucleotide.
  2. complex acids composed of nucleotides.

- **nucleoid**
  the region that harbours the chromosome of a prokaryotic cell. Unlike the eukaryotic nucleus, it is not bounded by a membrane.

- **nucleolus**
  a small, generally spherical body found within the nucleus of eukaryotic cells. The site of ribosomal RNA synthesis.

- **nutrient cycling**
  the transformation of chemical elements from inorganic form in the environment to organic form in organisms and, via decomposition, back to inorganic form.

- **nutrient**

1. any food or element an organism must take in to live, grow, or reproduce.
2. plant: an essential element in a particular ion or molecule that can be absorbed and used by the plant. For example, carbon, hydrogen, nitrogen, and phosphorus are essential elements; carbon dioxide, water, nitrate (NO_3) and phosphate (PO_4) are respective nutrients.
3. animal: Materials such as protein, vitamins and minerals that are required for growth, maintenance, and repair of the body and also materials such as carbohydrates that are required for energy.

- **obligate**

1. aerobe: an organism that requires oxygen for cellular respiration and cannot live without it.
2. anaerobe: an organism that cannot use oxygen and is poisoned by it.

- **obligate mutualism**

a condition in which a mutualistic relationship with other species is essential for a species to survive.

- **oceanic zone**

the region of ocean water lying over deep areas beyond the continental shelf.

- **octet rule**

the principle that chemical reactions are likely to result in the filling of the outermost shell of an atom (except hydrogen) with 8 electrons.

- **odour signature**

the odour produced by an individual organism.

- **oiling**

rubbing behaviour of birds serving to distribute oil from the oil gland at the base of the tail over the feathers; in particular rubbing the back of the head on the back.

- **old growth**

old forests often containing several canopy layers,

variety in tree sizes and species, decadent old trees

and standing and dead woody material.

- **old-growth forest**
virgin and old, second growth forests containing trees that are often hundreds, sometimes thousands, of years old. These are the richest forest biomes with the widest arrays of niche microhabitats and the broadest biodiversity, especially in the tropics.

- **old factory**
referring or relating to the sense of smell.

- **oligophagous**
consuming a small range of types of food items.

- **oligotrophic**
a nutrient-poor, clear, deep lake with minimum phytoplankton.

- **ommatidium**
discrete optical unit of the insect compound eye, with a lense focusing light from a small angle of the visual field onto a small group of photoreceptors.

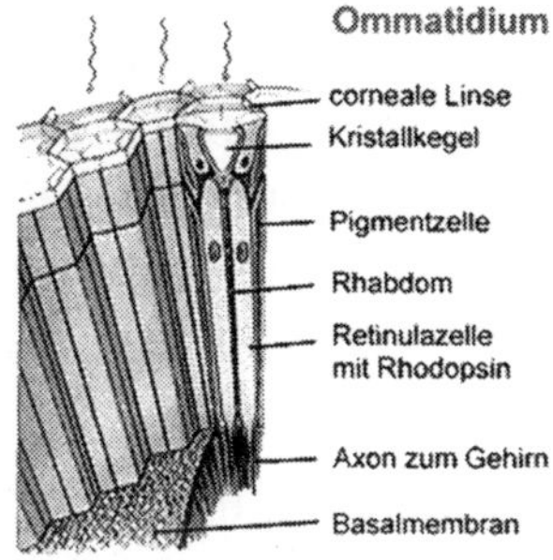

- **oncogene**
a gene found in viruses or as part of the normal genome that is involved in triggering cancerous characteristics.

- **on-the-dot sampling**
a system of sampling in which the behaviour is recorded at a specific time interval.

- **ontogenesis**
individual development.

- **ontogenetic**
occurring during the course of an organism's development.

- **ontogeny**
the development of characters such as behaviour over the life of the individual.

- **oogamy**
  a condition in which male and female gametes differ, such that a small, flagellated sperm fertilises a large, non-motile egg.

- **oogenesis**
  the process in the ovary that results in the production of female gametes.

- **oophagy**
  the eating of eggs; egg cannibalism.

- **oparin**
  a Russian scientist who made contributions to ideas concerning prebiotic chemical evolution following research on polynucleotide interactions and coacervate formation.

- **open circulatory system**
  an arrangement of internal transport in which blood bathes the organs directly and there is no distinction between blood and interstitial fluid.

Circulatory System Variations

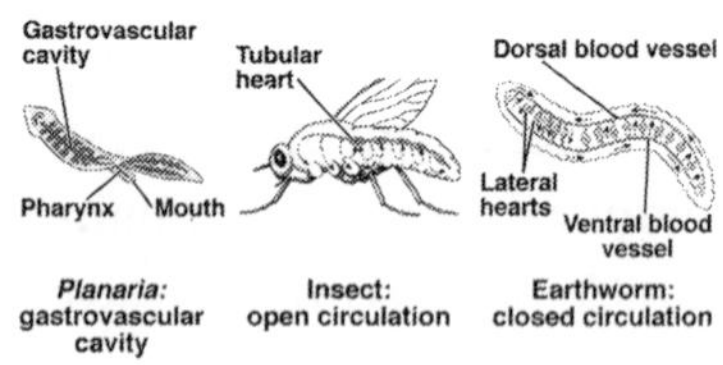

- **open loop**
  any type of circuit, electrical or neural, in which there is no feedback to help regulate the performance of the circuit.

- **open system**
  a system, such as a living organism, in which both matter and energy are

exchanged between the system and the environment.

- **open-field test**
  an animal is placed in an open arena and its activity is measured.

- **open-loop behaviour**
  behaviour that is performed without feedback control, characteristic of behaviours where speed or response time is crucial, such as escape responses.

- **operant behaviour**
  the totality of operant responses in the behavioural repertoire of the animal.

- **operant conditioning**
  the experimental procedure of presenting the animal with a reinforcing stimulus immediately following the occurrence of a given response. Conditioning is said to occur if, and only if, the response then increases in rate of occurrence, magnitude, or relative frequency, or decreases in latency, as a consequence of this operation.

- **optomotor response**
  a feedback system by which animals adjust their position with respect to a moving background. When one rotates a drum around an animal, the individual responds by turning its body in the opposite direction as though it is attempting to stabilise the image of the cylinder's stripes on its retinae.

- **ordinal data**
  variables categorised into mutually exclusive groups arranged with reference to their magnitude.

- **ordination**
  mathematical system for categorising communities on a graph so that those that are most similar in species composition appear closest together.

- **organ**
  a specialised centre of body function composed of several different types of tissues.

- **organ of corti**
  the actual hearing of the vertebrate ear, located in the floor of the cochlear canal in the inner ear; contains the receptor cells (hair cells) of the ear.

• **organelle**
a structure that is found within a cell and carries out one or more specific functions within the cell.

• **organic**
pertaining to living organisms in general, to compounds formed by living organisms and to the chemistry of compounds containing carbon.

• **organic soil**
soil at least partly derived from living matter, such as decayed plant material.

• **organism**
any living creature.

• **organisational effect**
the effect of certain hormones in shaping the development of the proximate mechanisms that underlie a behavioural ability.

• **organogenesis**
an early period of rapid embryonic development in which the organs take form from the primary germ layers.

• **orgasm**
rhythmic, involuntary contractions of certain reproductive structures in both sexes during the sexual response cycle.

• **orientation**
the perception of one's position in space. Primary orientation controls the body posture whereas secondary orientation controls the response to a stimulus from the environment.

• **orientation flight**
after leaving the nest the animal flies or moves around the area and learns the landmarks or other cues of its environment so that it can return to the home base.

- **orogeny**

the process of the formation of mountains.

- **orthogenesis**

the discredited theory that evolution is directed toward some ultimate goal.

- **orthokinesis**

1. refers to kinesis orientation behaviour in which the animal moves at different rates under different conditions.
2. a change in the rate of random movement of an organism in which the rate of location varies with the intensity of the stimulus.

- **orthologous gene**

duplicate corresponding genes at the same loci in different species.

- **orthoselection**

selection exerted upon a lineage in the same direction for a substantial segment of geological time.

- **ORV**

Off-Road Vehicles, such as motor cycles, 4-wheel drive vehicles and 4-wheelers.

- **osmoconformer**

an animal that does not actively adjust its internal osmolarity because it is isosmotic with its environment.

- **osmolarity**

solute concentration expressed as molarity.

- **osmoregulation**

1. regulation of the salt concentration in cells and body fluids.
2. adaptations to control the water balance in organisms living in hyperosmotic, hypoosmotic, or terrestrial environments.

- **osmoregulator**

an animal whose body fluids have a different osmolarity than the environment, and that must either discharge excess water if it lives in a hypoosmotic environment or take in water if it inhabits a hyperosmotic environment.

- **osmosis**

1. diffusion of water across a semi-permeable membrane.
2. the diffusion of water across a selectively permeable membrane.

- **osmotic pressure**
  the tendency of water to move across a semi-permeable membrane into a solution.

- **osteocyte**
  a cell located in lacunae of vertebrate bone, that deposits a matrix of collagen and calcium phosphate that forms a mineralised connective tissue.

- **ostracoderm**
  an extinct agnathan; a fish-like creature encased in an armour of bony plates.

- **outbreeding**
  in which genetically dissimilar organisms mate with each other.

- **ovarian cycle**
  the cyclic recurrence of the follicular phase, ovulation and the luteal phase in the mammalian ovary, regulated by hormones.

- **ovary**
  1. in flowers, the portion of a carpel in which the egg-containing ovules develop.
  2. in animals, the structure that produces female gametes and reproductive hormones.

- **overburden**
  layer of soil and rock overlying a mineral deposit, removed during surface mining.

- **overcompensating density dependence**
  density dependence which is so intense that increases in initial density lead to reductions in final density.

- **overconsumption**
  situation in which some people consume much more than they need at the expense of those who can not meet their basic needs- and at the expense of earth's present and future life-support systems for humans and other forms of life.

- **overcrowding**
  an excessively high spatial density of animals due to some abnormal environment. The consequences depend on its intensity and duration.

- **oxidising**
  a term pertaining to a process or environment in which a chemical element or ion is oxidised via the loss of electrons.
- **oxidising agent**
  the electron acceptor in a redox reaction.
- **p**
  the frequency of an allele at one locus in a population.
- **p site**
  peptidyl-tRNA site; the binding site on a ribosome that holds the tRNA carrying a growing polypeptide chain.
- **p53**
  a nuclear phosphoprotein which acts as a tumour suppressor. The gene for p53 is located on the short arm of chromosome 17. Neoplasms can be induced by the inactivation of p53 by mutation, or by virtue of interaction with oncogene products of DNA tumour viruses. Therefore, p53 is considered to be a tumour suppressor. In addition, human cancers are associated with mutations in the p53 gene.
- **pacemaker**
  internal timekeeping mechanism capable of generating or coordinating circadian rhythms.
- **pacemaker neuron**
  neuron that in the absence of any input spontaneously generates individual spikes or bursts of spikes.
- **paternity assurance or certainty**
  refers to the probability that the offspring a male is helping to rear are his own. Specific adaptations evolve such as mate guarding to increase paternity certainty.
- **pathogen**
  a micro organism or virus that causes disease.
- **pattern formation**
  the ordering of cells into specific three-dimensional structures, an essential part of shaping an organism and its individual parts during development.
- **pawing**
  a scooping leg movement, generally on the ground.
- **PCR**
  Polymerase Chain Reaction is a technique for iso-

lating and amplifying any specifically desired DNA sequence. The reaction is facilitated by a heat-stable DNA polymerase (e.g., Taq, which is obtained from a thermophilic bacterium) that can withstand the many cycles of heating and cooling involved in the technique.

- **peck order**
  social hierarchy in domestic birds.

- **pecking**
  making contact with an object or the body of another organism using the tip of the beak.

- **pedigree**
  a family tree describing the occurrence of heritable characters in parents and offspring across as many generations as possible.

- **pelagic**
  a term pertaining to the open ocean environment or to the organisms inhabiting that environment.

- **pelagic zone**
  the area of the ocean past the continental shelf, with areas of open water often reaching to very great depths.

- **penetrance**
  a measure of the proportion of individuals homozygous for a gene that show its phenotypic effect.

- **peptide**
  the connecting group in a protein or polypeptide chain. The linkage results from a dehydration reaction involving two amino acids.

- **peptidoglycan**
  the cell wall material of many prokaryotes, consisting of a single molecule that surrounds the entire cell.

- **peramorphosis**
  development is accelerated or lengthened relative to the ancestor so that maturity occurs at an older developmental stage.

- **pesticides**

chemicals that kill or inhibit the growth of organisms that people consider undesirable. Fungicides (which kill fungi), herbicides (which kill plants), and insecticides (which kill insects) are types of pesticides.

- **petiole**

1. the stalk of a leaf.
2. the stalk of a leaf, which joins the blade to a node of the stem.

- **pH**

a scale of acidity (1 - 7) or alkalinity (7 - 14) derived from the logarithm of the concentration of hydrogen ions (10^{-1} - 10^{-14}).

- **phage**

a virus that infects bacteria; also called a bacteriophage.

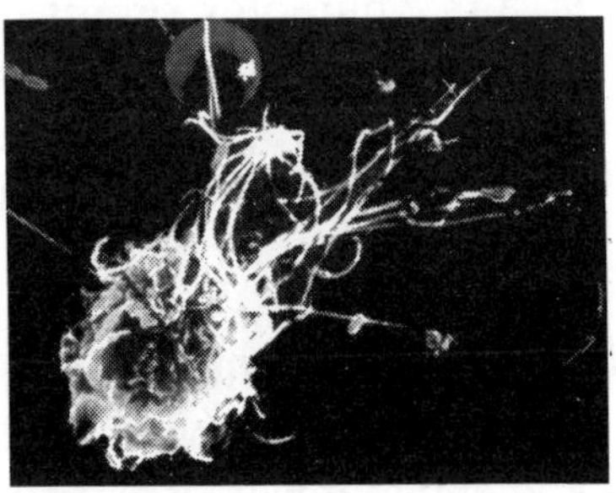

- **phagocyte**

white blood-corpuscle capable of destroying harmful bacteria.

- **phagocytosis**

a type of endocytosis involving large, particulate substances.

- **phanerozoic**

the more recent eon of earth history composed of the Paleozoic, Mesozoic and Cenozoic eras.

- **pharynx**

an area in the vertebrate throat where air and food passages cross; in flatworms, the muscular tube that protrudes from the ventral side of the worm and ends in the mouth.

- **phase-resetting**

experiments which reveal whether or not a particular neuron is involved in the clock mechanism that determines the timing of cycles in a rhythmically repeating activity. Injection of a brief pulse of current into a neuron should reset the rhythm by advancing or delaying the time of subsequent bursts of spikes in the interneurons and motor neurons depending on the phase of the cycle in which the stimulus current is delivered.

- **phasic receptor**
type of receptor cell in which adaptation is so rapid that the response of the cell to the stimulus declines to zero very quickly. See also **Tonic Receptor**

- **phenetic**
pertaining to phenotypic similarity; used especially to describe taxonomic relationships based on phenotypic similarity in contrast to phyletic relationships based on recency of common ancestor.

- **phenocopy**
a modification of the phenotype owing to special environmental conditions which resembles a change of the phenotype known to be caused by a mutation.

- **phenogram**
a phylogenetic tree that emphasizes similarity. See **cladogram**.

- **phenology**
1. strictly the study of periodic biological events; in practice often applied to periodic phenomena themselves, such as the lifetime pattern in an organism of growth, development and reproduction in relation to the seasons.
2. study of the temporal aspects of recurrent natural phenomena and their relation to weather and climate.
3. the timing of recurring natural phenomena, and its study. In ethological contexts phenology refers mainly to the timing of breeding seasons, especially in relation to cyclic changes in weather conditions.

- **phenotype**
1. a visible, or otherwise measurable, physical or biochemical characteristic of an organism, resulting from the interaction between the genotype and the environment.
2. any character or set of characters of an organism; the morphological, physiological, biochemical, behavioural and other properties of an organism, manifested throughout its life, that develop through action of genes and environment; or any subset of such properties, especially those affected by a particular allele or other portion of the genotype.

- **phenotype matching**

a means of kin recognition by which an animal uses its own characteristics or those of kin with which it has associated to distinguish relatives from nonrelatives. It enables an animal to tell a relative that it has not encountered before from a nonrelated stranger. In the case of sweat bees (Lasioglossum zephyrum) it has been shown that a female guarding a nest will admit strangers as long as they are of the same family as the bees with which the female was raised, even if these were not her own siblings. A proximate mechanism of kin discrimination in which an individual's behaviour toward another is based on how similar they are in appearance, odour, or some other phenotypic trait.

- **phenotypic plasticity**

the ability to develop different phenotypes under different conditions.

- **pheromone**

1. a chemical substance, usually a glandular secretion, that is used in communication within a species. One individual releases the material as a signal and another responds after detecting it. A chemical substance, usually a glandular secretion, that is used in communication within a species. One individual releases the material as a signal and another responds after detecting it. Specific chemicals that function in intraspecific communication. Also sometimes ectohormone or social hormone. Pheromones are hormone-like substances produced by glands that, in contrast to true endocrine glands, secrete to the exterior of the body. They thus affect not the individuals producing them but other individuals of the same species.

2. a small, volatile chemical signal that functions in communication between animals and acts much like a hormone in influencing physiology and behaviour.

- **philopatry**

1. the tendency to return to the natal site. The tendency of an individual to return to

(or stay in) and breed in its home area (birthplace or other adopted locality).
2. limited dispersal where average propagule moves less than 10 homeranges away from natal site.

- **phloem**
  the portion of the vascular system in plants consisting of living cells arranged into elongated tubes that transport sugar and other organic nutrients throughout the plant.
- **phobia**
  an excessive, persistent and apparently unsubstantiated fear specific to a given stimulus, situation or circumstance.
- **phosphate**
  a functional group important in energy transfer.

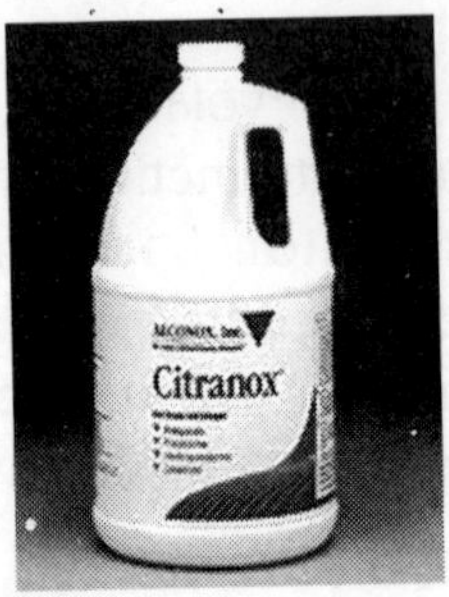

- **phospholipid**
  a molecule that contains phosphorus and is soluble in organic solvents. A primary component in cellular membranes.
- **phosphorus cycle**
  cyclic movement of phosphorus, in varying chemical forms, from the environment to organisms and then back to the environment.
- **photic zone**
  the narrow top slice of the ocean where light is adequate to support photosynthesis.
- **photoautotroph**
  an organism that harnesses light energy to drive the synthesis or organic compounds from carbon dioxide.
- **photoheterotroph**
  an organism that uses light to generate ATP but that must obtain carbon in organic form.
- **photon**
  discrete packet of light energy; the amount of which depends on the colour or wavelength of light.

- **photoperiod**
  length of the period of daylight each day.
- **photoperiodism**
  a physiological response to day length, such as flowering in plants.
- **photophosphorylation**
  the process of generating ATP from ADP and phosphate by means of a protonmotive force generated by the thylakoid membrane of the chloroplast during the light reactions of photosynthesis.
- **photoreceptor**
  1. receptors sensitive to light.
  2. receptor cell specialised to respond to light energy.
- **photorespiration**
  a metabolic pathway that consumes oxygen, evolves carbon dioxide, generates no ATP, and decreases photosynthetic output; generally occurs on hot, dry, bright days, when stomata close and the oxygen concentration in the leaf exceeds that of carbon dioxide.
- **photosynthate**
  the energy-rich organic molecules produced during photosynthesis.
- **photosynthesis**
  the conversion of light energy into chemical energy that is stored in glucose or other organic compounds; occurs in plants, algae and certain prokaryotes.

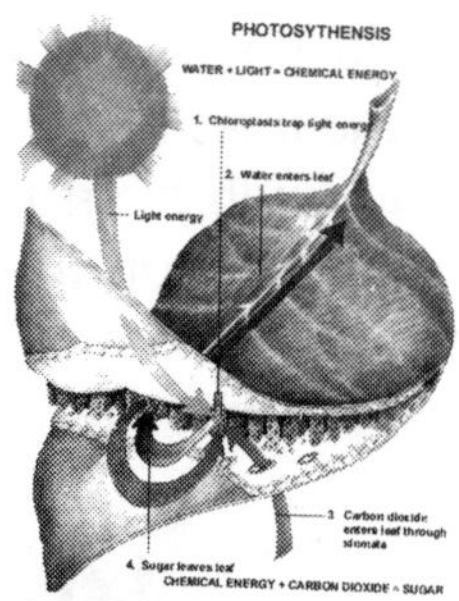

- **Photosynthetically Active Radiation (PAR)**
  those wavelengths in the spectrum of radiation that are effective in photosynthesis.

• **photosystem**
the light-harvesting unit in photosynthesis, located on the thylakoid membrane of the chloroplast and consisting of the antenna complex, the reaction-centre chlorophyll a, and the primary electron acceptor. There are two systems which absorb light best at different wavelengths.

• **phototaxis**
an orientation response with respect to light.

• **phototropism**
growth of a plant shoot toward or away from light.

• **phyletic**
transformation of an unbranched lineage of organisms sometimes to a state different enough from the ancestral population to justify renaming it as a new species; also called anagenesis.

• **phytoplankton**
small, usually microscopic, aquatic organims (e.g., bacteria, protists, plants) capable of photosynthesis. The more general term 'plankton' is a general designation for various drifting microscopic aquatic organisms in the upper regions of aquatic ecosystems both photosynthetic and non-photosynthetic.

• **pica**
abnormal appetite for unusual and often inappropriate food, e.g. dirt, hair, faeces.

• **piloerection**
in many mammals the erection of the fur or hair in connection with threat possibly as a means of effecting a more impressive appearance.

• **pilot fish**
fish that accompany sharks, mantas and serve as cleaners for them or they may

just feed on wasted food. They are often seen swimming near the dorsal fin of sharks.

- **piloting**
  orientation by the use of local landmarks.

- **placenta**
  a structure in the pregnant uterus for nourishing a viviparous foetus with the mother's blood supply; formed from the uterine lining and embryonic membranes.

- **placental mammal**
  a member of a group of mammals, including humans, whose young complete their embryonic development in the uterus, joined to the mother by a placenta.

- **placoderm**
  a member of an extinct class of fishlike vertebrates that had jaws and were enclosed in a tough, outer armour.

- **plankton**
  mostly microscopic organisms that drift passively or swim weakly near the surface of oceans, ponds and lakes.

- **plasma**
  the liquid matrix of blood in which the cells are suspended.

- **plasma cells**
  derivatives of B cells that secrete antibodies.

- **plasma membrane**
  the membrane at the boundary of every cell that acts as a selective barrier that regulates the cell's chemical composition.

- **plasmagenes**
  hereditary factors located in the cytoplasm; a cytoplasmic factor.

- **plasmid**
  extrachromosomal, autonomously replicating, circular DNA segment.

- **plural breeding**
  a social system in which two or more females (monogamous or not) breed in the same communal social unit. The term does not apply to colonial species unless communal units exist within the colony. See **singular breeding**.

- **pluripotent stem cell**
  a cell within bone marrow that is a progenitor for any kind of blood cell.

- **poaching**
hunting, trapping or fishing illegally.

- **pogonophoran**
a marine invertebrate of the phylum Pogonophora.

- **poikilotherm**
an organism whose body temperature is strongly correlated with that of its external environment.

- **point mutation**
1. a mutation within a gene, such as the changes in a codon, addition or loss of a base, etc.
2. a change in the chromosome at a single nucleotide within a gene.

- **poisson distribution**
a mathematical distribution used to describe or test for a random distribution or process.

- **polarity**
of a character, whether postulated homologies are ancestral or derived.

- **pole timber**
trees at least 5 inches in diameter, but smaller than the minimum size for saw timber.

- **pole/sapling**
the stage of forest succession in which trees are between 3 and 7 inches in diameter and are the dominant vegetation.

- **pollen grain**
an immature male gametophyte that develops within the anthers of stamens in a flower.

- **pollination**
the placement of pollen onto the stigma of a carpel by wind or animals. Pollination is a prerequisite to fertilisation.

- **pollutant**
a particular chemical or form of energy that can adversely affect the health, survival, or activities of humans or other living organisms.

- **polyandry**
refers to a mating system in which females have more than one mate during a breeding season whereas the males normally have only one. A mating system characterised by a behavioural bond between one breeding female and two or more males breeding with her.

- **polycentric distribution**
  the presence of a population, species or other taxonomic group in several widely separated places.
- **polyclimax theory**
  the idea that succession leads to one of a variety of climaxes, depending on local environmental conditions.
- **polyethism**
  behavioural polymorphism, e.g. age polyethism is the regular changing of roles by the individuals of an insect colony as they age.
- **polygenic character**
  a trait, the variation in which is based wholly or in part on allelic variation at several or many loci.
- **polygenic- inheritance**
  an additive effect of two or more gene loci on a single phenotypic character.
- **polygeny**
  the capacity of several genes to influence the development of a trait.
- **polygynandry**
  a mating system characterised by behavioural bonds between two or more breeding females and two or more males breeding with them.
- **polygyny**
  a polygamous mating system involving one male and many females.
- **polymer**
  a large molecule made up of similar or identical monomeric subunits.
- **polymerase**
  enzymes that catalyse the assembly of nucleotides into RNA and of deoxynucleotides into DNA.
- **polymerisation**
  the formation of large organic molecules from many like subunits.
- **polymorphic**
  referring to a population in which two or more morphs are present in readily noticeable frequencies.
- **polymorphism**
  the existence within a species or population of different forms of individuals, beyond those that are the result simply of recurrent mutation.
- **polynomial**
  a taxonomic designation consisting of more than two

terms; this method preceded the binomial nomenclature system.

- **polyp**
  the sessile variant of the cnidarian body plan. The motile phase is the medusa.

- **polypeptide**
  a polymer of many amino acids linked by peptide bonds.

- **polyphagous**
  consuming a wide range of types of food items.

- **polyphyletic**
  a taxonomic grouping that excludes the stem ancestor and excludes some of the descendants. Species are grouped based on convergent similarities.

- **polyploidy**
  the condition in which the number of chromosome sets in a cell is a higher multiple of the haploid number than two (which is diploidy). See **allopolyploidy, autopolyploidy**.

- **polyploidy**
  a chromosomal alteration in which the organism possesses more than two complete chromosome sets.

- **polyribosome**
  an aggregation of several ribosomes attached to one messenger RNA.

- **polysaccharide**
  a carbohydrate polymer made up of a chain of monosaccharides, e.g. starch, cellulose.

- **polytocous**
  referring to animals that typically deliver several offspring at each parturition.

- **polytopy**
  geographic variation in which each of one or more distinctive forms is found in each of several separate localities, between which other forms are distributed.

- **polytypic species**
  a species consisting of two or more subspecies.

- **polytypy**
  the existence of named geographic races or subspecies within a species.

- **population**
  1. a group of individuals of one species in an area, though the size and nature of the area is defined, often arbitrarily, for the purposes

of the study being undertaken.

2. a group of conspecific organisms that occupy a more or less well-defined geographic area and exhibit reproductive continuity from one generation to the next; it is generally presumed that ecological and reproductive interactions are more frequent among the individuals of a population than between them and the members of other populations of the same species.

- **population change**
an increase or decrease in the size of a population. It is equal to (births + immigration) - (deaths + emigration).

- **population cycle**
changes in the numbers of individuals in a population which repeatedly oscillate between periods of high and low density.

- **population density**
the numbers in a population per unit area, or sometimes 'per unit volume', 'per leaf' or whatever seems appropriate.

- **population dispersion**
general pattern in which the members of a population are arranged throughout its habitat.

- **population distribution**
variation of population density over a particular geographical area. For example, a country has a high population density in its urban areas and a much lower population density in rural areas.

- **population dynamics**
the variations in time and space in the sizes and densities of populations.

- **population ecology**
the study of the variations in time and space in the sizes and densities of populations and of the factors causing those variations.

- **population fitness**
the average of fitness of a set of genotypes belonging to the same population.

- **population fluctuations**
variations over time in the size of a population.

- **population pyramid**
a means of illustrating the age structure of a population diagrammatically, by plac-

ing the youngest age class at the base and stacking successive age classes above it.

- **population regulation**
a tendency in a population for some factor to cause density to increase when it is low and to decrease when it is high.

- **Population Vulnerability Analysis (PVA)**
an analysis, generally applied to populations or species in danger of extinction, of the population's chances of extinction.

- **population**
a group within a single species, the individuals of which can and do freely interbreed. Breeding between populations of the same species is less common because of differences in location, culture, nationality and so on.

- **position effect**
a difference in the phenotypic expression of a gene associated with a change in its location on the chromosomes.

- **position information**
signals, to which genes regulating development respond, indicating a cell's location relative to other cells in an embryonic structure.

- **positive feedback**
a physiological control mechanism in which a change in some variable triggers mechanisms that amplify the change.

- **poxvirus**
DNA viruses that are among the largest and most complex of all viruses.

- **practice**
the observed occurrence of a specified part of behaviour one or more times. The greater the number of occurrences, the greater the amount of practice.

- **pragmatics**
in communication the study of the relations of signals to their effects. The 'meaning' of communication.

- **Prairie grassland**
a local, North American, name for the temperate grassland biome.

- **praise**
a type of positive reinforcer given for a specific behaviour.

• **preadaptation**
the coincidental possession of traits favourable to a new environment before that environment is actually occupied, as air-breathing apparatus possessed by some aquatic animals prior to their becoming terrestrial. Does not imply adaptation for future use.

• **pseudoextinction**
the disappearance of a lineage from the fossil record due to its evolution into a new form rather than to its termination.

• **pseudofemale behaviour**
behaviour by males that appears to be similar to that of females.

• **pseudogene**
1. a sequence that looks like a gene but does not function as one; it appears to have no phenotype and could be the vestigial remains of a gene.
2. a noncoding DNA segment with sequences similar to functional genes but lacking the sites necessary for gene expression.

• **pseudo-interference**
a pattern of declining predator consumption rate with increasing predator density, reminiscent of the effects of mutual interference, but resulting from the aggregative response of the predator.

• **pseudomale behaviour**
behaviour by females that appears to be similar to that of males of the same species.

• **pseudopodium**
a cellular extension of amoeboid cells used in moving and feeding.

• **pseudopregnancy**
a condition that can occur in female mammals when mating fails to cause conception. The stimulation of copulation gives rise to the same hormonal changes that result from pregnancy and hence the other physiological changes that go with that condition, such as cessation of ovarian cycling.

• **pseudospecies**
a species that is derived through anagenic change, i.e., over time there has been sufficient change for it to be called a species by taxonomists.

- **psychology**
  the study of mental life and behaviour.

- **psychophysics**
  use of behavioural assays to establish sensory capabilities of an individual.

- **pteridophytes**
  a division of the plant kingdom comprising ferns, horsetails, clubmosses and their allies. Plants with true stems, leaves and roots (diploid) reproducing by spores and alternating with a free-living inconspicuous sexual (haploid) generation (prothalli).

- **public domain**
  the territory ceded to the Federal government by the original thirteen states, plus additions by treaty, cession and purchase.

- **public goods**
  goods that may be enjoyed by any number of people without affecting other peoples' enjoyment. For example, an aesthetic view is a pure public good. No matter how many people enjoy the view, others can also enjoy it.

- **public involvement**
  the use of appropriate procedures to inform the public, obtain early and continuing public participation, and consider the views of interested parties in planning and decision making.

- **public land**
  land for which title and control rests with a government—Federal, state, regional, county or municipal.

- **punctuated equilibria**
  most major evolutionary transitions and changes are temporally associated with speciation. There are periods of rapid speciation and rapid evolution with intervening periods of very slow rates of evolution.

- **punctuated evolution**
  see **punctuated equilibria**.

- **punishment**
  the procedure of following the occurrence of a conditioned response with an aversive stimulus.

- **pupillary response**
  the expansion and contraction of the pupils serve to regulate the amount of light

entering the eye. This mechanism is widespread among mammals.

- **pure strategy**
  when all individuals of the population have the same strategy.

- **purine**
  one of two families of nitrogenous bases found in nucleotides consisting of two members: adenine and guanine.

- **puzzle box**
  a general term used for any boxlike apparatus used to study learning, memory or other mental abilities.

- **pyramid of biomass**
  diagram representing the biomass (total dry weight of living organisms) that can be supported at each trophic level in a food web. The bottom of the pyramid is comprised of *primary producers* , while the peak of the pyramid is topped by one (or at most a small handful) *apex predator* . Humans are abnormal in that we cross all ecosystems and biomass pyramids, and in almost every one (excepting the polar caps and deepest of oceanic environments) we are the dominant apex predator.

- **pyramid of energy flow**
  also called a *trophic pyramid.* Diagram representing the flow of energy through each trophic level in a food chain or food web. With each energy transfer, only a small part (typically 10%) of the usable energy entering one trophic level is transferred to the organisms at the next trophic level, with the remaining 90% lost as heat or expended in metabolic processes.

- **pyrimidine**
  one of two families of nitrogenous bases found in nucleotides consisting of three members: cytosine (C), thymine (T), and uracil (U).

- **q**
  the frequency of an allele in a population.

- **q_{10}**
  the quotient of two reaction rates at temperatures differing by 10°C.

- **quadrat**
  a sampling unit used to assess density of organisms.

- **quantal speciation**
  speciation in allopatry which is rapid due to founder and bottleneck effects. See **peripatric speciation**.

- **quantitative character**
  a heritable feature in a population that varies continuously as a result of environmental influences and the additive effect of two or more genes.

- **quantitative ethology**
  a term sometimes used for the quantitative recording and subsequent analysis of behaviour. It provides an important basis for the compilation of an ethogram. The quantitative treatment of characteristics that allow recognition of bonding within a social group is referred to as sociometry.

- **quantitative trait**
  a trait that varies continuously.

- **quantum evolution**
  very rapid evolutionary change, often assumed to be involved in the production of new higher taxa.

- **quasisocial**
  a social group in which members of the same generation use the same nest and cooperate in brood care but all individuals breed (social insects).

- **quaternary structure**
  the particular shape of a complex, aggregate protein, defined by the characteristic three-dimensional arrangement of its constituent subunits, each a polypeptide.

- **queen**
  the reproductively dominant individual in a eusocial colony.

- **queenright**
  a colony with a functional queen.

- **quiescent centre**
  a region located within the zone of cell division in plant roots, containing meristematic cells that divide very slowly.

- **r**
  1. (group) in chemistry a shorthand designation which represents a variable part of a molecule. For example, the part of an amino

acid other than the amino and carboxyl groups.
2. (r- selection) the concept that in some populations, a high reproductive rate is the chief determinant of life history.

- ***r* selection**
selection of life-history traits which promote an ability to multiply rapidly in numbers - the traits being, broadly, small size, precocious reproduction, similarity, a large reproductive allocation and the production of many but small offspring.

- **rabbit-fox principle**
the notion that in arms races (such as between predator and prey) there may be unequal selective pressures on the two parties (in this case the rabbit is running for its life whereas the fox is running for its dinner which means that selection is stronger on the rabbit) and thus may win the arms race.

- **race**
a poorly defined term for a set of populations occupying a particular region which differ in one or more characteristics from other populations; sometimes referring to local variations or polymorphisms within a population and sometimes equivalent to a subspecies.

- **radial**
1. (cleavage) Embryonic development in deuterostomes in which the planes of cell division which produce the blastula are either parallel or perpendicular to the polar axis.
2. (symmetry) A body shape characterised by parts which radiate outward from the centre.

- **radiata**
members of the radially symmetric animal phyla (cnidarians and ctenophores).

- **radiative evolution**
an evolutionary process that involves the branching of an evolutionary lineage and so leads to greater systematic diversity.

- **radicle**
an embryonic root of a plant.

- **radioactive**
1. having a nucleus that gives off particles and en-

ergy; characteristic of unstable isotopes of a chemical element.
2. (dating) a method of determining the age of fossils and rocks using half-lives of radioactive isotopes.

- **ramet**
an offshoot or module formed by vegetative growth in some plants and modular invertebrates that is actually or potentially independent physiologically, e.g. the runners of the strawberry, the tubers of the potato, the polyps on a colonial hydroid.

- **random distribution**
lacking pattern or order. The result of (or indistinguishable from the consequence of) chance events.

- **random fixation**
the loss of an allele (and fixation of the other allele) in a population owing to accidents of sampling (genetic drift).

- **range**
land on which the principle natural plant cover is composed of native grasses, forbs and shrubs that are valuable as forage for livestock and big game.

- **range management**
the art and science of planning and directing range use intended to yield the sustained maximum animal production and perpetuation of the natural resources.

- **range of variability**
the components of healthy ecosystems fluctuate over time. The range of sustainable conditions in an ecosystem is determined by time, processes (such as fire), native species, and the land itself. For instance, ecosystems that have a 10 year fire cycle have a narrower range of variation than ecosystems with 200-300 year fire cycle. Past management has placed some ecosystems outside their range of variability. Future management should move such ecosystems back toward their natural, sustainable range of variation.

- **rank - abundance diagram**
a graphical plot of differential abundances of species in a community.

- **rank mimicry**
imitation by lower-ranking individuals of the behavioural characteristics of higher-ranking individuals. In primates it may play a role independent of rank. An acoustic example of rank mimicry has been noted in some birds, in which young males selectively learn the songs of higher-ranking males. So far this tendency has been established only in polygynous and promiscuous species, in which it probably aids the mimicking males in their efforts to attract females.

- **rank order**
Ordered distribution of 'rights and obligations'-precedence and sub ordinance-within a group of animals living together. The stronger or more experienced individuals stand higher in the order and consequently enjoy certain privileges, such as first access to resources.

- **ranked preference**
a preference exhibited by a consumer between food items all of which can be classified on the same simple scale (usually energy value).

- **rape**
forced copulation.

- **raptor**
a bird of prey, such as a eagle or hawk.

- **rare enemy effect**
the intensity of selection depends on the number of predators, and if predation is not common, selection may not favour a phenotype that reduces predation.

- **rare male effect**
the situation in which females prefer the male that is different from the others, i.e. female preference for a phenotype depends on its frequency in the population (males are preferred when rare). This is a frequency-dependent effect. When the mating success of a male depends on the frequency of his phenotype in the population.

- **rate of evolution**
rate at which a population adapts to a change in the environment.

- **rate of response**
the number of responses occurring in unit time.

- **reaction**

a chemical change in which a substance decomposes, combines with other substances, or interchanges constituents with other substances. (centre) The location of one or a pair of specialised chlorophyll a molecules in the pigment assembly system of the light reactions of photosynthesis.

- **reaction norm**

the normal pattern of reactions of an animal to directional stimuli in which the form of the movement remains independent of the respective stimulus.

- **reading**

one-way linear process by which nucleotide sequences are decoded, for example, by protein-synthesising systems.

- **realised niche**

1. that portion of its potential (fundamental) niche occupied by a species when competitors or predators are present.
2. set of resources and physical habitats actually used by individuals of a species in an area.
3. the environmental resources that an organism actually uses; the niche that an organism actually occupies.

- **realm**

a region of the Earth that harbours similar groups of species based on a shared biogeographical history. For example, while primates are found in many parts of the world, all monkeys found with prehensile tails are only found in the Neotropic realm. The Earth is made up of eight realms.

- **rebound spike**

spike in the neuron triggered by the release from a hyperpolarising event.

- **recapitulation**

the repetition of ancestral adult stages in embryonic or juvenile stages of descendants. The ontogenetic passage of an organism's features through stages that resemble the adult features of its phylogenetic ancestors.

- **receptive field**

1. for any unit in a sensory system, the area from which

it receives input that changes its activity. The effects can be either an increase in the activity of the unit (positive input) or a decrease (negative input). For example, most of the axons (neuron fibres) in the optic nerves of a cat have receptive fields at specific spots on the retina; these consist of an excitatory centre with an encircling inhibitory ring. This arrangement is a consequence of the fact that a retinal receptor cell excites the optic nerve neurons immediately underlying it but inhibits those with which it connects (via intermediate cells) in the surrounding area. This process of lateral inhibition has the effect of enhancing contrast in perception of the edges of retinal images. Units have been discovered in the visual systems of frogs that have receptive fields tuned to the visual patterns presented by flying insects. These 'bug detectors' provide the kind of stimulus filtering that releasing mechanisms were postulated to provide.

2. in sensory systems, the specific region of a sensory surface that when stimulated causes a change in activity of a neuron.

- **receptor**
  sensory site.

- **receptor cell**
  peripherally located neuron, often anaxonal, that can detect and respond to a sensory stimulus. May also be used generically to mean any neuron that can respond to a sensory stimulus.

- **receptor potential**
  electrical potential that appears in sensory neurons when a stimulus is applied.

- **recessive**
  1. of an allele, the phenotypic trait is not revealed when heterozygous, only when homozygous.
  2. in a heterozygote, the allele that is completely masked in the phenotype.

- **recharge**
  the addition of water to ground water by natural or artificial processes.

- **reciprocal altruism**
  altruistic behaviour between unrelated individu-

als; believed to produce some benefit to the altruistic individual in the future when the current beneficiary reciprocates.

- **reciprocal predation**

an interaction between two species (or individuals) in which each preys upon the other, so that the interaction is, in essence, competitive.

- **recognition**

1. retentive discrimination. The ability to differentiate between two categories of stimuli, but as properties of objects, places, or individuals rather than mere sensory discrimination (see **Perception**). In most cases recognition is a result of learning (as in individual recognition), but there are exceptions, such as species recognition.

2. (concept of species) A definition of species based on mate-recognition mechanisms; assumes that reproductive adaptations of a species consist of a set of features that maximise successful mating with members of the same population; an alternative to the biological species concept.

- **recombinant**

1. an offspring whose phenotype differs from that of the parents.

2. (DNA) a technique in which gene segments from different sources are recombined in vitro and transferred into cells, where the DNA may be expressed.

- **recombination**

1. the formation in offspring of combinations of genes not present in either parent. This results from the assortment of chromosomes and their genes during the production of gametes and the subsequent union of different sorts of gametes at fertilisation.

2. the formation of new combinations of genetic material by crossing-over during synapsis in meiosis.

- **recruitment**

additions to a population, either through birth or immigration, or, in the case of net recruitment, the differences between such additions and the losses resulting from death or emigration.

- **recruitment behaviour**

behaviour of worker social insects that attracts other

individuals of the colony to a new food source, nest site or intruder in the nest.

- **red queen hypothesis**
  the idea that animals are in a continual evolutionary arms race with other members of their species and with other species and that they are never able to get ahead because of the co-evolutionary response.

- **redirected behaviour**
  behaviour directed at an inappropriate or substitute stimulus object or individual. Any response elicited by a stimulus and, ordinarily, topographically directed toward that stimulus, which, in the presence of a drive-conflict, is topographically directed at a part of the environment that otherwise would not elicit it.

- **redox**
  the term applied to chemical reactions which involve loss of one or more electrons by one molecule (oxidation) and simultaneous gain by another (reduction).

- **reducing**
  a term pertaining to a process or environment in which a chemical element or ion is reduced via the gain of electron. (opposite of oxidising).

- **reduction**
  a phenomenon whereby a substance gains one or more electrons. The gain is accompanied by an increase in free energy.

- **reductionism**
  1. the belief that complex phenomena can be entirely explained by reducing them to the smallest possible component parts and explaining these.
  2. explanation of the molar in terms of the molecular; accounting for surface appearances by showing them to be consequences of an underlying structure or process on a more fundamental level. In behaviour reductionism usually takes the form of showing how behaviour is consequent on physiological mechanisms, which in turn can be understood in terms of their physics and chemistry. Reductionism is typically atomistic and deterministic.

- **redundancy**

in information theory redundancy exists when more signs are used to encode a message than would be sufficient to do so. It is deliberately included by the communications engineer when noise is likely to disturb signals in transmission. Redundancy is measured as the ratio of the difference between the most efficient coding and the observed coding to the most efficient coding. In a language such as English, redundancy enables us to read a text full of typographical errors: differences in the probabilities of letter sequences and of word combinations, syntactic constraint, and semantic connection combine to provide clues to what the words are supposed to be in most of the garbled cases. Most animal communication is repetitive and structured in ways that make it highly redundant. The genetic code is redundant because many of the amino acids to which it refers are represented by more than one triplet sequence of bases.

- **reflex arc**

information comes into the animal through a specific receptor organ which responds to the stimulus, the information travels down an afferent pathway to the ganglion cells of the CNS; the efferent nerves carry the impulse to the effector organ which produces the appropriate response.

- **reflex sensitisation**

if, after repeated elicitation by its stimulus, a reflex response is then given in response to a previously neutral or much less effective stimulus with which the reflex stimulus has not been paired, reflex sensitisation is said to have occurred.

- **reflex**

1. an involuntary response of an organism to a stimulus in which the response proceeds in a clearly prescribed fashion or reflex arc.
2. any S-R correlation, whatever its history or probability of occurrence.
3. a S-R correlation which is observable in all members of

a species under a given set of conditions in which there have been experimental manipulations calculated to prevent learning. This is often further qualified as cases where change in strength are negligible with respect to drive-inducing operations.

- **reforestation**
  the restocking of an area with forest trees, by either natural or artificial means, such as planting.

- **refractory period**
  1. brief period following an action potential during which it is more difficult than normal or impossible to excite a neuron to generate another action potential.
  2. the short time immediately after an action potential in which the neuron cannot respond to another stimulus, owing to an increase in potassium permeability.

- **refugia**
  areas in which a species was confined during a period in its evolution and from which it later expands its range.

- **regeneration**
  the renewal of a tree crop by either natural or artificial means. The term is also used to refer to the young crop itself.

- **regression analysis**
  analysis of the mathematical relationship between two variables.

- **regular distribution**
  the arrangement of individuals with respect to each other which has some pattern or order that ensures that they are more widely separated from each other than would be expected by chance.

- **regulative- development**
  a pattern of development, such as that of a mammal, in which the early blastomeres retain the potential to form the entire animal.

- **regulatory sequence**
  a DNA sequence involved in regulating the expression of the gene (e.g. promoters and operators).

- **reinforcement**
  1. the operation of presenting to the animal in operant conditioning, after it has made a response (and there-

fore contingent on its occurrence), a reinforçing stimulus or of withdrawing a negative reinforcing stimulus.

2. in classical conditioning, the operation of presenting, contiguously in time, a conditioned stimulus and an unconditioned stimulus.

- **reinforcement: primary**

the use in reinforcement of a stimulus that is a reinforcing stimulus to any animal of a given species without special training.

- **reinforcement: secondary**

when a stimulus, initially ineffective, becomes a reinforcing stimulus after being presented to an animal contiguously in time with a reinforcing stimulus, it is termed a secondary or conditioned reinforcing stimulus and the operation of presenting it is termed secondary reinforcement.

- **reintroduce**

to return members of a species to their historical range. This strategy is sometimes used when a species has become locally extinct or if its population is threatened.

- **relative**

1. (abundance) the frequency of individuals of a species in a biological community; also called species equitability.

2. (fitness) the contribution of one genotype to the next generation compared to that of alternative genotypes for the same locus.

- **relative humidity**

very roughly, the dampness of the air; more correctly, the percentage saturation of the air with water vapour; better still, the mass of water vapour per unit volume of air as a percentage of the same measure for saturated air at the same temperature.

- **release cutting**

removal of competing vegetation to allow desired tree species to grow.

- **releaser**

1. in ethology body markings and postures or behaviour patterns which are adaptations for communication.

2. releaser is used as a synonym for sign stimulus.

3. a sign stimulus produced by the physical structure or

behaviour of an animal that releases some particular species-specific response or responses of a conspecific.

- **releasing hormone**
  a hormone produced by neurosecretory cells in the hypothalamus of the vertebrate brain that stimulates or inhibits the secretion of hormones by the anterior pituitary.

- **rhinovirus**
  viruses that replicate in cells of the upper respiratory tract and cause the common cold.

- **rhizosphere**
  1. the surface and immediate neighbourhood of a root which provides a specialised environment for micro organisms.
  2. the term is also used to define the micro flora that lives in this region.

- **rhodopsin**
  a visual pigment consisting of retinal and opsin. When rhodopsin absorbs light, the retinal changes shape and dissociates from the opsin, after which it is converted back to its original form.

- **rhythms**
  regularly recurring patterns of behaviour. Circannual rhythms are those occurring in phase with the tides. Lunar rhythms are those occurring in phase with the moon. Circadian rhythms are those associated with the day-night cycle.

- **ribosomal RNA**
  the most abundant type of RNA. Together with proteins, it forms the structure of ribosomes that coordinates the sequential coupling of tRNA molecules to the series of mRNA codons.

- **ribosome**
  1. granules associated with the boundary of the endoplasmic reticulum; the presumptive sites of protein synthesis.
  2. a small organelle that is the site of protein synthesis.

- **ribozyme**
  an enzymatic RNA molecule; catalyses reactions during RNA splicing.

- **richness**
  the number of species in a biological community.

- **ring-barking**
  the removal of bark and the immediately underlying living tissues from a ring around the stem or trunk of a tree or shrub. This severs the route of transport of leaf products down to the roots.

- **riparian**
  pertaining to the banks of a river or other body of fresh water.

- **riparian area**
  the area along a watercourse or around a lake or pond.

- **riparian ecosystem**
  the ecosystems around or next to water areas that support unique vegetation and animal communities as a result of the influence of water.

- **risk**
  the chance of incurring a loss in reproductive success.

- **risk-sensitive or risk-averse behaviour**
  behaviour in which options are chosen with respect to the variance in the reward or punishment. The demonstration of risk sensitivity requires that basis for choice is not the option with the highest mean. This is done by making the means of the options equal and the variances unequal or by coupling the desired and chosen variance with the undesired mean.

- **ritualisation**
  1. process by which a functional behaviour pattern or structure is transformed into a communication signal.
  2. the term is used generally to refer to the modification of a behavioural pattern to serve a communicative function. Ritualisation is usually associated with an action becoming more stereotyped, less variable between conspecifics and of a typical intensity.
  3. an evolutionary process by which behaviour patterns become modified into displays.
  4. the evolutionary modification of a behaviour pattern that turns it into a signal used in communication or improves its efficiency as a signal.

- **ritualised fight**
  innate fighting pattern which is directed against conspecifics and which is stereotyped and appears to proceed according to set rules and does not normally result in damage despite the fact that the opponents may bear dangerous weapons.

- **ritualised response.**
  1. a response that appears in most members of a species, that is relatively invariant in topography, and that is typically a social releaser.
  2. a response that appears in the repertoire of the members of several different but related species and that usually varies in stereotypy from species to species.

- **Ribo-Nucleic Acid (RNA)**
  giant molecules of various types, some of which, in most organisms, copy the genetic code from the DNA and carry it to the sites of protein synthesis (messenger RNA) and some aid in the organisation of the synthesis (transfer RNA).
  1. (ribonucleic acid) a nucleic acid containing ribose as the sugar within the nucleotides. The genetic storage material of some viruses.
  2. (polymerase) the enzyme which links ribonucleotides together during transcription.
  3. (processing) modification of RNA within the nucleus of eukaryotes.

- **Robert Hooke**
  coined the term 'cell' to describe the structures he saw while examining a piece of cork using a microscope.

- **rod cell**
  one of two kinds of photoreceptors in the vertebrate retina; sensitive to black and white and enables night vision.

- **rolling**
  an activity characterised by lying down and rolling onto the back accompanied by rubbing movement against the ground. Considered to be a comfort movement.

- **roost**
  refers to the place where birds spend the night. A communal roost is where a

number of individuals come together during the night in a social group.

- **root hair**
  a tiny projection growing just behind the root tips of plants, increasing surface area for absorption of water and minerals.

- **root pressure**
  the upward push of water within the stele of vascular plants, caused by active pumping of minerals into the xylem by root cells.

- **ROS**
  Recreation Opportunity Spectrum. The land classification system that categorises land by its setting and the probable recreation experiences and activities it affords.

- **rotation**
  the number of years required to establish and grow timber crops to a specified condition of maturity.

- **rough ER**
  that portion of the Endoplasmic Reticulum studded with ribosomes.

- **round dance**
  with rapid rocking steps, the forager bee moves in circles which are often so tight that only one comb cell lies inside. The round dance signifies that a rich source of food lies in the vicinity of the hive, to a distance of 25 m.

- **roundwood**
  timber and fuelwood prepared in the round state, such as house logs and telephone poles.

- **r-selection**
  selection that favours organisms producing large numbers of offspring each with a relatively low probability of survival; often associated with low competitive ability, short lifespan, rapid development, early reproduction and good dispersal ability.

- **rubisco**
  ribulose carboxylase, the enzyme that catalyses the first step (the addition of CO_2 to RuBP, or ribulose bisphosphate) of the Calvin cycle.

- **ruderal**
  a plant of waste places, usually associated with human

disturbance. The word distinguishes this group of plants from 'weeds' which are plants that are a nuisance to human activities - the ruderal is not necessarily a nuisance.

- **ruffling**

erection of feathers. Without doubt the primary function of feather ruffling is temperature regulation. In this connection three feather positions have been distinguished: ruffling, or extreme erection, with feather tips separated from one another, thus opening a passage for air to get to the skin for cooling; fluffing, in which erection stops short of separation of the feather tips and so increases the thickness of the insulating layer for heat retention; and sleeking, which contrasts with erection in that the feathers are depressed against the skin surface, thus reducing the thickness of insulation and reducing heat retention. Feather erection may be restricted to certain parts of the body, depending upon the conditions. For example, the feathers around the brood patches are ruffled as part of the sequence of settling movements with which a bird lowers itself onto its eggs to incubate them.

Special forms of feather ruffling have frequently evolved for the secondary function of display, often with associated modification of the shape or coloration of particular feathers, as in crests and plumes, the capes of pheasants, and the ruffs of ruffs. Such ritualised ruffling is especially prominent in the displays of threat behaviour and courtship, where they may function as releasers necessary for the conduct of social interaction. For example, in many of the species in which allopreening occurs, a slight ruffling of the plumage can indicate where on its body the recipient of the attention would like the preening directed.

- **ruminant**

1. herbivorous mammals such as cows that chew the cud and have complex stomachs containing micro organisms that break down the cellulose in plant material.

2. an animal, such as a cow or a sheep, with an elaborate, multicompartmentalised stomach specialised for a herbivorous diet.

• **runaway process**
refers to the situation in which females choose males on the basis of heritable, favourable characters and in this way the genes for the female preference and the genes for the favourable male character end up in the same individuals. When this happens the female preference genes increase in the population by virtue of their association with the favourable male character. This can result in the evolution of extravagant male characters. A form of sexual selection that occurs when female mating preferences for certain male attributes create a positive feedback loop favouring males with these attributes and females that prefer them.

• **run-off**
the portion of precipitation that flows over the land surface or in open channels.

• **runoff**
surface water effluent (usually from precipitation but may be from human activities such as irrigation) that moves too quickly to be absorbed into the ground. It flows down contour gradients to enter stream and river systems, carrying with it anything light enough to be borne in the volume of water, which may be light after a small rain or tremendous in the wake of a storm, when even large boulders and trees get swept up in the runoff. When runoff travels over *deforested* or unplanted agricultural lands, it carries away large quantities of *topsoil* . Runoff from agricultural areas often carries heavy doses of *biocides* , fertilisers, and other nutrients, which can lead to *eutrophication* when introduced into aquatic systems.

• **rut**
in mammals a periodically occurring state of sexual and mating behaviour controlled by male sex hormones.

- **s**
  the phase during interphase of the cell cycle when DNA is replicated.

- **s.**
  1. selection coefficient;
  2. speciation rate.

- **SA node**
  the pacemaker of the heart, located in the wall of the right atrium.

- **sacrifice area/site**
  in range management, a site allowed to be overgrazed to obtain efficient overall use of the management area. In cultural resource management, it may refer to a site intentionally sacrificed to extensive public use in order to preserve the larger cultural area.

- **salinisation**
  accumulation of salts in soil that can eventually make the soil unable to support plant growth.

- **salt**
  a substance composed of one or more ions.

- **salt pan**
  a basin or pool containing an accumulation of salty water or salt.

- **saltation**
  a discontinuous, major evolutionary change in phenotype occurring between two generations.

- **saltatory conduction**
  1. action potentials move down nerves at an increased rate due to the presence of myelin and the fact that only intermittent regions of the neuron are depolarised.
  2. rapid transmission of a nerve impulse along an axon resulting from the action potential jumping from one node of Ranvier to another, skipping the myelin-sheathed regions of membrane.

- **sample**
  a subset of a population; a group of items drawn by some stated procedure from a population and from which one hopes to learn certain things about the population.

- **sampling error**
  statistical deviation from the previous genetic composition of a population resulting from random factors in the production or selection

of the next generation; in reference to genetic drift in small populations.

- **sanitation salvage**
  the removal of dead, damaged or susceptible trees primarily to prevent the spread of pests or disease and promote forest health.
- **sapling**
  a loose term for a young tree more than a few feet tall and an inch or so in diameter that is typically growing vigorously.
- **saprobe**
  an organism that acts as a decomposer by absorbing nutrients from dead organic matter.
- **saprophyte**
  an organism that carries out external digestion of non-living organic matter and absorbs the products across the plasma membrane of its cells (e.g. fungi).
- **sarcomere**
  the fundamental repeating unit of striated muscle, delimited by the Z lines.
- **sarcoplasmic reticulum**
  a modified form of endoplasmic reticulum in striated muscle cells that stores calcium used to trigger contraction during stimulation.
- **satellites**
  periperial males. Males that remain near mated or paired males who are displaying conspicuously or maintaining a territory. In this way these satellite males may gain access to the female.
- **satiation**
  a state of motivation, usually, but not always referred to in the context of feeding. When feeding undisturbed from a source the animal spontaneously ceases.
- **satiety**
  a state of motivation in which an animal stops performing a behaviour even though the external stimulus situation is still present. Often used in the context of sexual behaviour.
- **saturated**
  a fatty acid in which all carbons in the hydrocarbon tail are connected by single bonds, thus maximising the number of hydrogen atoms attached to the carbon skeleton.

- **savanna**

1. type of grassland with widely spaced trees and a blanket of grasses that require a lot of light. Rainfall usually occurs in the warmer, summer months with a dry period of between two to eight months. Fires are typical across savannas during drier months and occur at intervals from one to 50 years.

2. a tropical grassland biome with scattered individual trees, large herbivores, and three distinct seasons based primarily on rainfall, maintained by occasional fires and drought.

- **saw timber**

trees that are 9 inches in diameter at breast height or larger that can be made into lumber.

- **scale**

in ecosystem management, it refers to the degree of resolution at which ecosystems are observed and measured.

- **scan sampling**

a whole group of subjects is rapidly scanned or censured at regular intervals and the behaviour of each individual at that instant is recorded.

- **scanning electron microscopy**

an instrument which uses a focused electron beam to study the surface features of small objects.

- **scansorial**

refers to climbing or adapted for climbing.

- **scatter hoarding**

the caching of food in a dispersed manner through the environment in contrast to larder hoarding.

- **scent marking**

the deposition of chemical substances on the substrate. They are usually secreted by specific scent organs and function in communication, particularly territory defence.

- **Schleiden**

contributed to the formulation of the cell theory.

- **schooling**

the formation of a social group in fish, often tightly coordinated in their behaviour.

- **schwann**

1. contributed to the formulation of the cell theory.

2. a type of supporting cell enclosing the axons of many neurons and forming an insulating layer called the myelin sheath.

- **scientific method**
  a systematic approach to problem solving characterised by hypothesis testing.

- **scientific truth**
  accepted as true because available evidence supports it and all efforts to falsify it have failed; relies on our understanding of natural phenomena; remains open to further evaluation.

- **sclereid**
  a short, irregular sclerenchyma cell in nutshells and seed coats and in the parenchyma of some plants.

- **sclerenchyma**
  a rigid, supportive plant cell type usually lacking protoplasts and possessing thick secondary walls strengthened by lignin.

- **scoping**
  the ongoing process to determine public opinion, receive comments and suggestions and determine issues during the environmental analysis process. It may involve public meetings, telephone conversations or letters.

- **score-keeping mutualism**
  a type of mutualism that depends upon repeated exchange of favours and in which each participant adjusts its behaviour in response to partial or total nonreciprocation by the other. Two of a family of such strategies for the prisoner's dilemma are Tit-for-tat and Judge. See also **mutualism and reciprocity.**

- **scout bees**
  bees that search for new supplies of food or a new nest site; when returning home they transmit their news to their hive mates with a dance.

- **scramble competition**
  the most extreme form of overcompensating density dependence in the effect of intraspecific competition on survivorship where all competing individuals are so adversely affected that none of them survive.

- **scramble competition polygyny**
a mating system in which males are nonterritorial and seek scattered, receptive females.

- **scratching behaviour**
birds scratch on the substrate by using first one and then the other leg as in chickens, called alternate scratching. Other species such as thrushes and finches spring forward with both legs at the same time pulling the claws backwards over the surface, called hop scratching.

- **scutellum**
also called cotyledon; thin seed leaf with large surface area to absorb nutrients during germination; found in monocots such as corn.

- **search image**
1. in the search for food or prey, an animal develops through trial and error a certain memory of the goal object. From this behaviour results a searching image which produces selective searching. This leads to a change in the ability of the animal to locate particular types of food. This is particularly important when searching for cryptic prey. A hypothetical mental picture of a prey item used by a predator to search specifically for a cryptic, common, edible prey.
2. restriction of an animal's interest to a single class of object as a consequence of focused attention on particular stimuli.

- **searching efficiency**
the instantaneous probability that a given predator will consume a given prey (also called 'the attack rate').

- **second growth**
forest growth that was established after some kind of interference with the previous forest crop, such as cutting, fire or insect attack.

- **second law of thermodynamics**
in any conversion of heat energy to useful work, some of the initial energy input is always degraded to lower quality, more dispersed, less useful energy — usually low-temperature heat that flows into the environment; every energy system has leaks and looses energy or heat to attenuation.

- **second messenger**
  a chemical signal, such as calcium ions or cyclic AMP, that relays a hormonal message from a cell's surface to its interior.

- **secondary**
  1. (compounds) synthesised by prey species for defence (through diversion of metabolites from food species).
  2. (structure) the localised, folding of a polypeptide of a protein due to hydrogen-bond formation.
  3. (succession) recolonisation of an area following a disturbance such as fire.

- **secondary growth**
  growth in plants due to the action of cambium layers.

- **secondary productivity**
  the rate at which biomass is produced per unit area by heterotrophic organisms.

- **sedimentary rock**
  rock formed from sand and mud that once settled in layers on the bottom of seas, lakes, and marshes; often rich in fossils.

- **seed**
  an adaptation for terrestrial plants consisting of an embryo packaged along with a store of food within a resistant coat.

- **seed bank**
  the population of viable dormant seeds that accumulates in and on soil and in sediments under water.

- **seed tree harvest**
  removal of the mature timber crop from an area in one cut, except for a certain number of seed bearers.

- **segregation**
  the separation of homologous alleles at meiosis.

- **segregation distorter**
  when the separation of homologous alleles at meiosis is not even. Any of several mechanisms by which there is a non-random distribution of chromosomes in to the daughter cells at meiosis. See **meiotic drive.**

- **selection**
  1. non-random differential survival or reproduction of classes of phenotypically different entities. See **Natural selection, artificial selection, sexual section**.
  2. (coefficient) The difference between two fitness

values, representing a relative measure of selection against an inferior genotype.
3. (sexual) Selection based on variation in secondary sex characteristics, leading to enhancement of sexual dimorphism.

- **selection coefficient**
  the measure of the disadvantage in survival a given genotype holds relative to that of the most fit genotype in the population; The relative rate by which the frequency of a given genotype is reduced each generation by selection.

- **selectionist**
  differences in the amount of genetic variation in a population are due to the action of natural selection; also, generally the viewpoint that the observed characteristics of animals are due to the action of natural selection (adaptations) rather than being due to other mechanisms such as mutation and genetic drift.

- **selective**
  1. (agent) the environmental characteristic which, by exhibiting change, serves to provide for change in a population through natural selection.
  2. (permeability) a property of biological membranes that allows some substances to cross more easily than others.

- **selective agent**
  the factor responsible for a particular selective pressure.

- **selective breeding**
  artificial selection.

- **selective pressure**
  1. a force acting on populations that determines that some individuals leave more descendants (or genes) than others to subsequent generations (and so gives direction to the process of evolution.)
  2. any feature of the environment that results in natural selection; for example, food shortage, the activity of a predator, or competition from other members of the same sex for a mate can cause individuals of different genotypes to survive to different ages, to reproduce at different rates or both.

- **selfish herd**
  a group of individuals each one of which uses the others as living shields against predators.

- **self-limitation**
  a process where intraspecific competition leads to a reduction in reproduction and/or survival at higher densities.

- **self-thinning**
  the progressive decline in density which accompanies and interacts with the increasing size of individuals in a population of growing individuals.

- **semantics**
  in communication, the study of relations of signals to their referents. The messages of communication.

- **semelparity**
  where organisms produce all of their offspring in a single reproductive event over one relatively short period.

- **semen**
  the fluid that is ejaculated by the male during orgasm; contains sperm and secretions from several glands of the reproductive tract.

- **semicircular canals**
  a three-part chamber of the inner ear that functions in maintaining equilibrium.

- **semilethal**
  a recessive lethal gene which lacks complete penetrance and expressivity, that, kills less than 100% of the homozygotes before reaching sexual maturity.

- **semilunar valve**
  a valve located at the two exits of the heart, where the aorta leaves the left ventricle and the pulmonary artery leaves the right ventricle.

- **seminal vesicle**
  a part of the male reproductive tract that stores sperm in invertebrates and produces semen in vertebrates.

- **seminiferous tubules**
  highly coiled tubes in the testes in which sperm are produced.

- **semiotics**
  the study of communication.

- **semispecies**
  1. one of several groups of populations that are par-

tially but not entirely reproductively isolated from each other by biological factors (isolating mechanisms); populations that border between subspecies and typical species.
2. populations which are reproductively isolated in some areas of their range, but intrograde in others.

- **senescence**
aging; a progression of irreversible change in a living organism, eventually leading to death.

- **sensation**
1. neuronal activity resulting from the transduction of stimulus energy into electrical activity.
2. an impulse sent to the brain from activated receptors and sensory neurons.

- **sense organ**
an organ specialised for the reception of sensory stimulation.

- **sensitive period**
period of time during which a developing animal is especially sensitive to particular types of experience.

- **sensitive species**
plant or animal species which are susceptible to habitat changes or impacts from activities.

- **sensitisation**
enhancement of response to most stimuli following exposure to a significant signal.; related terms: pseudoconditioning; example: We blink every time a puff of air is blown onto our cornea. A subsequent sound will also produce an eye blink even though UCS (air puff) and CS (tone) had never been paired with each other.

- **sensory**
1. (neuron) receives information from the environment and transmits the signals to the CNS.
2. (nervous system = afferent) part of the peripheral nervous system that transmits signals to the CNS from sensory receptors.
3. (receptor) a structure that transduces environmental stimuli into action potentials.
4. refers to the sense organs.

- **sensory adaptation**
  decreased signalling of a peripheral sense organ with continued exposure to a stimulus.

- **sensory exploitation**
  in communication when a signaller is able to tap into the pre-existing biases of the sensory system of the receiver with the consequence that its message has more impact than it otherwise would.

- **sensory filtering**
  receptors responding only to very specific types of disturbance.

- **sensory neuron**
  neuron that carries sensory information into the central nervous system. See **Interneuron and Motor neuron.**

- **sensory receptor**
  cell endowed with the ability to respond to a specific stimulus (i.e., absorb a specific kind of stimulus energy).

- **sensory selectivity**
  subset of stimuli, which an animal detects and responds to. See also **Umwelt**.

- **sensory transduction**
  chain of physiological reactions in a sensory receptor, which convert sensory stimuli into electrical impulses; see also **Frequency code.**

- **sentinel behaviour**
  in many species one or several individuals sit in a conspicuous location and warn other members of the group of impending danger, often, associated with loud calls or visual displays.

- **sepal**
  a whorl of modified leaves in angiosperms that encloses and protects the flower bud before it opens.

- **separate nesting**
  a system of parental care in plural-breeding birds in which each breeding female has her own nest, as opposed to joint nesting.

- **separation call**
  a vocalisation made when an individual is removed from its group or becomes separated from conspecifics or other social partners.

- **sequential hermaphrodite**
  refers to a hermaphroditic species in which the organism is only one sex at a time, being first male (or female) and then turning into the other sex later.

- **sequential patterning**
  the pattern of producing different action patterns, for example the bout structure or the kinematic structure of behaviour.

- **seral**
  the stage of succession of a plant or animal community that is transitional. If left alone, the seral stage will give way to another plant or animal community that represents a further stage of succession.

- **serial homology**
  a relationship among repeated, often differentiated, structures of a single organism, defined by their similarity of developmental origin; for example, the several legs and other appendages of an arthropod.

- **serial monogamy**
  refers to a mating system in which animals are monogamous for one breeding season but then normally change their mates before the next breeding season.

- **seroreversion**
  an unexplained phenomenon whereby an individual who, through serological testing, is determined to have been exposed to an antigen later tests negative for the same antigen. This has been seen in some HIV patients.

- **serotiny**
  the retention by a plant (usually by trees) of seeds in hard enclosing structures, e.g. ovaries or cones so that they are not dispersed and free to germinate until after some disaster, especially forest fire.

- **serpentine soil**
  soil formed by the weathering of serpentine rock which contains high concentrations of various heavy metals. Serpentine soils are commonly localised and bear a specialised flora of species tolerant of these metals and of locally specialized 'ecotypes' of species that are also found elsewhere.

- **serum**
  that part of the blood plasma that remains after clots have formed and been removed.

- **sessile organism**
  literally a 'seated' organism. One whose position is fixed in space except during a dispersal phase, e.g. a rooted plant, barnacles, mussels (Mytilus), corals.

- **set point**
  the threshold of the regulatory machinery of an organism to maintain homeostasis.

- **sex**
  1. (chromosomes) The pair of chromosomes that differentiate between and are responsible for determining the sex of an individual.
  2. (-linked genes) Genes located on one sex chromosome but not the other.

- **sex attractants**
  pheromones produced by individuals of one sex that attract the opposite sex.

- **sex linked**
  a gene carried by one of the sex chromosomes; it can be expressed phenotypically in both sexes.

- **sex ratio**
  refers to the ratio of males to females in a population. Primary sex ratio is measured at the end of parental care; functional sex ratio is measured at the time of breeding. See also **operational sex ratio.**

- **sex-limited**
  a trait which is found in only one sex.

- **sexual**
  1. (dimorphism) Polymorphism based on the distinction between secondary sex characteristics, leading to the enhancement of sexual dimorphism.
  2. (reproduction) Reproduction in which parents produce offspring that have unique combinations of genes inherited from the gametes of the two parents.

- **sexual behaviour**
  includes all behaviour leading to the fertilisation of eggs by sperm. Once fertilisation has taken place, there may be further sexual behaviour directed toward further fertilisations or there may be a switch to some form of parental care.

- **sexual cannibalism**
  the consuming of the mate during or shortly before or after copulation.

- **sexual dimorphism**
  differences in appearance between males and females.

- **sexual imprinting**
  a learning process occurring in young animals that determines what its preferences will be in pair formation.

- **sexual recombination**
  the process by which DNA is exchanged between homologous chromosomes by chromosome pairing and crossing-over at meiosis during gamete formation.

- **sexual reproduction**
  production of offspring whose genetic constitution is a mixture of two (potentially) genetically different gametes.

- **sexual selection**
  1. the differential ability of individuals of different genetic types to acquire mates. Sexual selection consists of epigamic or intersexual selection, based on mate choices made by males (male choice) or females (female choice) and intrasexual selection, based on competition for mates between members of the same sex.
  2. process by which changes in gene frequencies result from individuals that are better than others at either competing for or at attracting mates - i.e. the evolution of traits based on differences in mating success. See also **Natural Selection.**

- **shaking to death**
  vigorous side-to-side head shaking of prey used by candies and a few other species where death of prey is caused by neck dislocation or suffocation.

- **sham attack**
  a ritualised threat behaviour in which the feigned attack is stopped just short of the opponent.

- **shaping**
  see **approximation conditioning**.

- **shelter wood**
  a cutting method used in a more or less mature stand, designed to establish a new crop under the protection of the old.

- **shifting balance theory of evolution**
  evolution would proceed most rapidly if natural populations were subdivided into small partially isolated groups where there is some genetic drift but not enough to lead to fixation. This was followed by interdemic selection between the differentiated subpopulations. This occurred by the selective diffusion of characters from demes with the more advantageous genetic combinations.

- **shoot**
  the aerial portion of a plant body, consisting of stems, leaves and flower.

- **short-day plant**
  a plant that flowers only when the light period is shorter than a critical length.

- **showing the nest behaviour**
  a component of courtship behaviour in many birds and fish in which the male draws the female toward a potential nest or nesting hole.

- **shredders**
  aquatic animals that feed on coarse particles of organic matter.

- **sib mating**
  mating of progeny derived from the same parents.

- **siblicide**
  the killing of siblings.

- **sibling species**
  morphologically similar or identical populations that are reproductively isolated. Species that are difficult or impossible to distinguish by morphological characters.

- **sickle dance**
  refers to a bee dance which is intermediate between the round dance and the waggle dance and indicates a rich source of food between 25 and 100 m.

- **siderophore**
  molecular receptors that bind and transport iron.

- **sieve tube member**
  a chain of living cells that form sieve tubes in phloem.

- **sigmoid curve**
  an 'S-shaped' curve in which there is an initial acceleration phase followed by a subsequent deceleration phase leading to a plateau.

- **sign stimulus**

1. a specified part, or change in a part, of the environment correlated in an orderly manner with the occurrence of a species-specific response that is not a reflex response.
2. particular subset of features or qualities of a natural stimulus or complex environmental cues sufficient to elicit a behavioural response. (e.g., FAP) via a releasing mechanism The stimulus to which a specific reaction pattern is tuned. Examples include the red spot on the beak of a herring gull, which evokes a pecking response from nestlings (which in turn prompts the parent to regurgitate food for the young); the size, colour, and degree of speckling of eggs vary the effectiveness as sign stimuli for the egg-rolling behaviour; Toads have particular sensitivity to small objects (potential prey items) moving in their visual field, and also to large objects (potential predators) moving in their visual field. See also **Stimulus**.
3. an external sensory stimulus that triggers a fixed-action pattern.

- **signal**

1. in ethology any behaviour that conveys information from one individual or from one group to another regardless of whether it serves other functions as well.
2. physical coding of information (e.g., a message) capable of transmission through environment.

- **signal jump**

a conspicuous and common courtship display behaviour of fish especially Pomacentridae. The fish swims upward above the substrate, the anterior end pointing slightly downward. Then the fish drops swimming rapidly downwards toward the substrate, thus inscribing an arc.

- **signal-transduction pathway**

a mechanism linking a mechanical or chemical stimulus to a cellular response.

- **signature display**

the species-typical head bobbing display pattern of territorial lizards.

• **signature whistle**
the unique calls of individual organisms, especially dolphins.

• **silent mutation**
mutation in a codon or sequence that does not cause an amino acid change.

• **silurian**
a geological era lasting from approximately 438 to 408 million years ago.

• **silverback male**
dominant male in groups of gorillas often with a grey marking on the back.

• **silvicultural system**
the cultivation of forests; the result is a forest of a distinct form. Silvicultural systems are classified according to harvest and regeneration methods and the type of forest that results.

• **silviculture**
the art and science that promotes the growth of single trees and the forest as a biological unit.

• **simple diffusion**
a physicochemical phenomenon whereby atoms, molecules, and ions move passively according to electrochemical gradients.

• **simultaneous hermaphrodite**
refers to a hermaphroditic species, animals or plants, which are both sexes at the same time.

• **single tree selection**
see **individual tree selection**.

• **singular breeding**
a communal social system in which no more than one female breeds in each social unit, as opposed to plural breeding.

• **sister chromatids**
replicated forms of a chromosome joined together by the centromere and eventually separated during mitosis or meiosis II.

• **site preparation**
the general term for removing unwanted vegetation, slash, roots and stones from a site before reforestation. Naturally occurring wildfire, as well as prescribed fire can prepare a site for natural regeneration.

• **sitting**
body position in which the posterior of the body trunk

is in contact with the ground and supports most of the body weight.

- **size class**
  one of the three intervals of tree stem diameters used to classify timber in the Forest Plan data base. The size classes are: Seedling/Sapling (less than 5 inches in diameter); Pole Timber (5 to 7 inches in diameter); Sawtimber (greater than 7 inches in diameter).

- **skeletal muscle**
  striated muscle generally responsible for the voluntary movements of the body.

- **skidding**
  hauling logs by sliding, not on wheels, from stump to a collection point.

- **skimming**
  a short period of running on the water and vigorous flapping of the wings by some species of waterfowl prior to take-off from water surface.

- **skinner box**
  1. animal placed inside the box is rewarded with a small bit of food each time it makes the desired response, such as pressing a lever or pecking a key. A device outside the box records the animal's responses.
  2. a space enclosed by a floor and walls of one or another material through which the study animal cannot escape. It is provided with one or more objects, the movement of which will automatically deliver a reinforcing stimulus. The manipulantum may be a key (to be pecked by a pigeon), a bar (to be pressed by a rat), or a string (to be pulled by a rat), or a panel (to be pushed by a dog or monkey). Used in studies of operant conditioning.

- **skototaxis**
  the movement of motile organisms toward darkness.

- **skyline logging**
  a logging system used to remove timber from steep slopes. Logs are brought upslope on a suspended cable, or skyline. Since the weight of the log is completely or partially supported by the cable, there is little disturbance to soil or other vegetation.

- **slash**
  the residue left on the ground after timber cutting or left after a storm, fire, or other event. Slash includes unused logs, uprooted stumps, broken or uprooted stems, branches, bark, etc.

- **slash and burnagriculture**
  an agricultural system in which farmers periodically clear land for farming by cutting and burning patches of forest. Traditionally, patches used for agriculture were allowed to revert to forests for a number of years before being replanted, causing minimal impact. Today, however, intensive slash and burn agriculture damages many tropical forest ecosystems.

- **slavery**
  the relation in which workers of a parasitic ant species raids the nest of another species, captures brood, usually in the form of pupae, and rears them as nest mates. After emergence the slaves work within the host colony.

- **sleeking**
  see **ruffling**.

- **sleep**
  1. the recurring phases of inactivity in the circadian cycle of an animal. The preparatory movements and sleeping postures are species specific.
  2. a behavioural phenomenon characterised mainly by prolonged periods of immobility and an increased reluctance to respond to stimulation.

- **sliding filament model**
  the theory of muscle contraction, which states that in the sarcomere, thin (actin) filaments slide across thick filaments, shortening the sarcomere; shortening of all sarcomeres in a myofibril shortens the entire myofibril.

- **slump**
  a landslide where the underlying rock masses tilt back as they slide from a cliff or escarpment.

- **small game**
  birds and small animals normally hunted or trapped.

- **small nuclear ribonucleoproteins**
  small particles in the cell nucleus, composed of RNA and protein molecules;

functions are not fully understood, but some form parts of spliceosomes, active in RNA splicing.

- **smiling**
  1. an expression common to all humans serves as an appeasement display to divert aggressive behaviour or as a salutation.
  2. in any mammal a facial expression in which the corners of the lips are drawn back.
- **smooth**
  1. that portion of the endoplasmic reticulum that is free of ribosomes.
  2. (muscle) a type of muscle lacking the striations of skeletal and cardiac muscle because of the uniform distribution of myosin filaments in the cell.
- **snag**
  a standing dead tree. Snags are important as habitat for a variety of wildlife species and their prey.
- **sneaky copulation**
  matings that occur outside the pair bond or harem, often by young males that avoid detection by the paired or harem males. also called cleptogamy.
- **sneezing**
  a sudden, powerful expulsion of air through the nose and mouth induced as a reflex spasm.
- **sniffing**
  a series of brief inhalations of air occurring during olfactory investigation.
- **social behaviour**
  behaviour having to do with interactions between conspecifics.
- **social Darwinism**
  a political or economic doctrine based on the survival of the fittest, where the fittest are those individuals, companies, countries or social systems which can survive in the face of competition from other similar units.
- **social deprivation**
  see **isolation experiment**.
- **social facilitation**
  1. an increase in consumption rate with increasing consumer density, resulting, for instance, from an increased time available for feeding when less time is required for vigilance against predators.

2. the tendency for an action by some members of a group to be taken up by others.

- **social group**
a group that is formed by mutual attraction of the participants to one another. (contrast with aggregation).

- **social impact**
significant events change people's lives, culture and/ or their community. Social impact assessment identifies how changes to an environment will affect or have affected people.

- **social inhibition**
suppression of an actions by one individual through the influence of another.

- **social justice and social equity**
the principles of social justice are access and equity. For social justice to occur there must be equal access to services and decision-making processes within and between communities. Equity is the sharing of resources within and between communities. Recognising and opposing discrimination on the basis of age, sex, race, religion, political persuasion and so on is one way of working towards social justice.

- **social learning**
see **tradition**.

- **social organisation**
the totality of social relationships from simple cooperation between males and females and territoriality to complex societies.

- **social parasite**
a species that takes advantage of the communication signals within social insect nests for furthering its own fitness. For example, there are species of beetle that acquire the odours of an ant colony and thus are accepted into the nest where they feed on the ant larvae.

- **social status**
the rank in the social hierarchy attained by an individual as a consequence of its interactions with other members of the group.

- **social unit**
groups of two or more individuals that associate together and behave in some respects as a unit, especially in group territoriality.

- **sociality**
  the tendency of animals to form social groups.
- **socialisation**
  acquisition of social capability through experience of interaction with conspecific as result of growing up in a group.
- **society**
  group of individuals of the same species that is organized in a cooperative manner extending beyond sexual and parental care.
- **sociobiology**
  a discipline that uses evolutionary theory as the foundation for the study of social behaviour. [JA] The study of social behaviour from an evolutionary perspective; the application of animal studies to human behaviour.
- **sociogram**
  a graphic representation of social relationships or interaction frequencies in a group. Individuals are usually represented as labelled boxes or circles connected with arrows to indicate how often each one does what to whom in a sampling of the behaviour of a group. Such an information network can display, for example, the dominance interaction in a hierarchically organised group.
- **sociometry**
  quantitative description of the social relationships within a group, the behavior patterns of its members, and the organization of the group. One of its methods consists of measuring the distances between group members and the directions they face relative to one another. See **Attention Structure**.
- **sodium-potassium ATP**
  type of membrane protein that uses the energy of ATP to move sodium ions out of the cell and potassium ions in the opposite direction across the membrane, thereby establishing their concentration gradients and creating the membrane potential.
- **sodium-potassium pump**
  a transport protein in the plasma membrane that transports sodium out of and potassium into the cell

against their concentration gradients.

- **soil compaction**
  the reduction of soil volume. For instance, the weight of heavy equipment on soils can compact the soil and thereby change it in some ways, such as in its ability to absorb water.

- **soil erosion**
  the loss of topsoil through silt-laden runoff, strong winds, or other forces that transport soil away from its natural location.

- **soil productivity**
  the capacity of a soil to produce a specific crop. Productivity depends on adequate moisture and soil nutrients, as well as favourable climate.

- **soldier**
  in social insects the caste that defends the colony from attackers, usually provided with larger defensive organs.

- **solitary**
  a designation used in biology for species in which individuals form no enduring groups or pair bonds, but live most of their lives in a solitary state.

- **solute**
  a molecule dissolved in solution. As a general rule, solutes dissolve only in solutions of similar polarity- for example, glucose (polar) dissolves in water (also polar), but not in oil (nonpolar).

- **solution**
  a homogeneous, liquid mixture of two or more substances.

- **solvent**
  medium in which a substance is dissolved.

- **soma**
  cell body of a neuron.

- **somatic**
  1. (cell) any cell in a multicellular organism except a sperm or egg cell.
  2. (nervous system) the division of the motor nervous system of vertebrates composed of motor neurons that carry signals to skeletal muscles in response to external stimuli.

- **somatic cell nuclear transfer**
  somatic cells are cells other than germline cells (egg cells

and sperm cells). The transfer technique indicated involves the transfer of the nucleus of a somatic cell (which contains the genome for the individual) to an enucleated egg cell. The egg cell now has a new nucleus, and it is the genome in the new nucleus that determines the development of the egg cell.

- **somatic polymorphism**
  the presence on the same genetic individual (genet) of organs of two or more different forms, e.g. several different leaf shapes on the same plant, two distinct types of seed produced on the same flower or inflorescence, modules of very different form in the body of a scalp.

- **somatotopic**
  topographical organisation of the somatosensory system, in which there is a point-to-point mapping of the surface of the body onto the somatosensory cortex.

- **somite**
  a block of mesoderm along each side of a chordate embryo.

- **song imprinting**
  said to apply in bird species in which the song is learned during a more or less clearly defined sensitive period, after which there is little or no capability of learning new or additional patterns. Song acquisition of this sort thus shows the essential characteristics of other kinds of imprinting (a sensitive period and irreversibility). However, such song often begins to emerge some time after the end of the sensitive period, so the 'imprinting' applies more to acquisition of a template, to which the bird later matches its song production, than to production of the song itself.

- **song matching**
  when the song type of one bird is matched by another, usually a neighbour or rival male.

- **song repertoire**
  in the bioacoustics literature, an inventory of the song types or song phrases available to an individual. The term is used mainly of male songbirds.

- **song**

  1. the long territorial vocalis-ations of passerines.

  2. any of a number of long and complex auditory signals produced by organisms. In ethology and bioacoustics, usually sound production of a length and complexity that distinguishes it from shorter and simpler calls. However, songs and calls overlap so much in both length and complexity that the distinction is often hard to sustain, at least on morphological grounds.

- **sonogram**

  tape recordings of vocalisations can be run through a sound spectrograph which registers the amplitude, frequency and sound pressure and prints this information as a sound picture on a special kind of paper.

- **sound pressure level**

  pressure generated by a sound wave measured on a logarithmic scale and expressed relative to the human threshold at 1kHz.

- **sound wood**

  timber that is in solid, whole, good condition. Sound wood is free from damage, decay or defects.

- **space constant**

  distance along a membrane at which a constant voltage will decay to 37% of its original size. This effectively measures the distance over which a postsynaptic potential can spread.

- **spacing behaviour**

  a behaviour pattern that serves to maintain a particular inter-individual distance.

- **spallanzani**

  conducted early research on spontaneous generation.

- **spatial summation**

  1. in a neuron, summation of two or more inputs of the same type from different locations that occur at about the same time.

  2. an accumulative effect on the membrane potential of a postsynaptic cell, caused by several different synaptic terminals stimulating a postsynaptic cell simultaneously.

- **spawning**

  the release of gametes or eggs into the water.

- **specialist**
feeds on only one or a very few food items.

- **specialist species**
1. a species that has a narrow ecological niche. For example, they may be able to live in only one type of habitat, tolerate only specific environmental conditions, or eat only a few types of food. The panda is a specialist species because over 95 percent of its diet consists only of bamboo.
2. species with a narrow ecological niche. They may be able to live in only one type of habitat, tolerate only a narrow range of climatic or other environmental conditions, or they may use only one or a few types of food.

- **specialisation**
a contracting, over evolutionary or individual time, of an individual's or species ecological roles or niche.

- **specialised**
applying to an organism or character judged to be more changed from a presumed common ancestral state than another with which it is compared.

- **speciation**
1. species originate when two separately evolving subspecies diverge to the point where the differences between them amount to between-species differences. As with the general difficulty of defining species, so is it often hard to decide when this point has been reached. As a rule speciation involves a period of geographic isolation, during which the subspecies undergo differential adaptation consequent on the differences between their habitats. See **allopatric**. Whether speciation can also occur when the subspecies occupy a single region (so-called sympatric speciation) is a question extensively discussed in the literature; indisputable evidence of sympatric speciation is still wanting. It is at least feasible that a group's very strong fixation on a particular part of the habitat could lead to reproductive isolation from other groups of individuals, which could lead to species-generating divergence. This, of course, may also be viewed as allo-

patric speciation, the spatial separation simply being on a much smaller scale—microgeographical isolation, as it were.

2. if two allopatric populations resume contact and if interbreeding is still possible, the hybrids may prove less fit than the offspring from matings within each group; in that case selection will result in the erection of barriers to interbreeding. Behavioural mechanisms, such as increased selectivity during pair formation, frequently contribute to the establishment of such reproductive isolation.

3. speciation of a different sort is shown in collections of fossil material. Here a single lineage may change over time to such an extent that forms from later strata may be considered as representing a different species from those in earlier strata. The judgments are made on the basis of morphological criteria and breaks in the fossil record. A species is judged as having been transformed into a different species as a consequence of accumulated difference with the passage of time. This view of speciation represents it as a continuous process of microevolutionary changes within species. This view has been challenged by one that describes phylogenetic history as consisting of punctuated equilibria.

4. phylogenetic split in a group of related individuals to the point where they become different enough to be considered separate species (e.g., unable to produce viable offspring). See **sympatric speciation.**

5. the process by which one or more populations of a species become genetically different enough to form a new species. The process often requires populations to be isolated for a long period of time.

6. the origin of new species in evolution.

- **speciation rate**

the numbers of new species added per million years.

- **species diversity**

an index of community diversity that takes into account both species richness and the relative abundance of species.

- **species recognition**
  1. in ethological writing, an exchange of stimuli between the sexes that ensures that the partners are conspecifics, thus preventing the mating of individuals of different species. The concept thus applies to those elements of courtship behaviour that govern the meeting of opposite-sex individuals of the same species, which accordingly serve to effect reproductive isolation.
  2. the lower animals have mechanisms for species recognition that are just as effective as those of higher vertebrates, yet most ethologists would dispute or dismiss the idea that such animals have consciousness.

- **species replacement**
  when two species come in contact with one another (after a period of allopatry), one of the two may outcompete and eliminate the other from the region of overlap.

- **species richness**
  the number of species present in a community.

- **species selection**
  differential survival of species over evolutionary time.

- **species**
  1. the boundaries of this taxonomic level (the most precise in the hierarchical system of *binomial nomenclature*) are hotly debated among scientists and there is little real consensus about where to draw the lines between species, subspecies, morphs, races, variants, etc. In general, a species is a group of organisms that resemble one another in appearance, general behaviour, ecological niche, chemical makeup and processes, and genetic structure. Organisms that reproduce sexually are classified as members of the same species only if they can actually or potentially interbreed with one another and produce fertile offspring. It should be noted that some (though quite few) taxonomists believe the species

level of classification is frequently invalid and these scientists only recognise classifications down to the level of genus (again, these taxonomists represent a very small minority view).
2. a phenotypically distinct, reproductively isolated aggregate of interbreeding populations.
3. in the sense of biological species, the members in aggregate of a group of populations that interbreed or potentially interbreed with each other under natural conditions.
4. a basic taxonomic category to which individual specimens are assigned, which often but not always corresponds to the biological species.
5. a group of organisms that have a unique set of characteristics (like body shape and behaviour that distinguishes them from other organisms. If they reproduce, individuals within the same species can produce fertile offspring.
6. the basic unit of biological classification. Scientists refer to species using both their genus and species name. The house cat, for example, is called *Felis catus*.
7. a population or group of populations whose members have the potential to interbreed (biological species).
8. a theory maintaining that species living the longest and generating the greatest number of species determine the direction of major evolutionary trends.

- **species-area relationship**
  1. there is a correlation between the equilibrium number of species found on an island and the area of the island (m^2 or km^2); also thought to apply to mainland situations.
  2. a common pattern in which the number of species on islands decreases as island area decreases.

- **species-deletion stability**
  tendency in a model community for the remaining species to remain at locally stable equilibria after a species is made extinct.

- **species-specific behaviour**
  1. those behaviours shown by a great majority of mem-

bers of a species in the same or highly similar environments and under the same or highly similar conditions. 2. those behaviours, complex and relatively stereotyped, that appear in most members of a species under set and stable conditions without evident prior opportunity to learn.

- **species-specific response**

a response that is a part of species-specific behaviour.

- **specific action potential or SAP**

1. the potentiality or readiness of the whole animal for response to a given stimulus. 2. lorenzian concept of reaction specific energy is an internal activator of behaviour. It should be replaced by SAP which can actually be measured from the behaviour. An animal's SAP status is dependent on its readiness or tendency to perform a particular instinctive behavioural pattern in preference to all other behaviours. This tendency declines or ends after the consummatory act of the instinctive behaviour has been performed.

- **specific heat**

the amount of heat (in calories) required to raise the temperature of 1 gram of a substance primary C. The specific heat of water is 1 calorie per gram.

- **spectral sensitivity**

in the study of animals' visual capacities, the relationship between response threshold and light frequency. For example, the activity in an animal's optic nerve fibre can be recorded while shining light of different frequencies on its receptive field on the retina to find the minimum intensity of each frequency that will cause a change in the activity of the fibre. The result is usually a U-shaped curve, the low point of which indicates the part of the frequency spectrum to which the unit is most sensitive.

- **speculum**

1. a white or coloured patch on the body surface that contrasts with its background.

2. German ethologists use the term for the white area around and below the base

of the tail of many social mammals (ungulates, lagomorphs, primates), as well as for the often iridescent patches on the wings of birds, the denotation most common in English.
3. the specula of ducks serve as a conspicuous signal for group cohesion, and males show them off by special movements during courtship. In the history of ethology they were among the first examples of intraspecific releasers.

- **sperm competition**
since females often mate with more than one male, there is often competition among ejaculates within the reproductive tract of the female for fertilisations or between sperm of different males in external fertilising species. Specific behaviours and morphological structures have evolved to improve a male's chances at this sperm competition.

- **spermatogenesis**
the continuous production of mature sperm cells in the testis.

- **spermatozoon**
the male gamete; a haploid, usually small, flagellated cell.

- **sphincter**
a ring like valve, consisting of modified muscles in a muscular tube, such as a digestive tract; closes off the tube like a drawstring.

- **spike**
action potential, so called because it usually has a sharp, pointed appearance on an oscilloscope.

- **spike initiation zone**
region of an axon at which an action potential is normally initiated when the neuron is excited.

- **spindle**
1. the ellipsoidal mass of protein fibres (microtubules) visible during cell division and thought to play a role in chromosome movement and division of the cytoplasm.
2. an assemblage of microtubules that orchestrates chromosome movement during eukaryotic cell division.

- **spiracle**
a hole in the sides of insects through which the tracheal

respiratory system connects with the exterior and which can be opened and closed.

- **spiracles**
minute pores distributed over the surface of an insect's body that serve as entrances to the tracheal system.

- **spiral cleavage**
a type of embryonic development in protostomes, in which the planes of cell division that transform the zygote into a ball of cells occur obliquely to the polar axis, resulting in cells of each tier sitting in the grooves between cells of adjacent tiers.

- **spite/spiteful behaviour**
social behaviour in which there is a net loss to the donor and to the recipient.

- **spliceosome**
a complex assembly that interacts with the ends of an RNA intron in splicing RNA; releases an intron and joins two adjacent exons.

- **splicing**
1. gene splicing: manipulations, the object of which is to attach one DNA molecule to another;
2. RNA splicing: the removal of introns from mRNA precursors.

- **spontaneous activity**
many neurons fire continuously and have a resting level of spontaneous activity.

- **spontaneous behaviour**
behaviour that occurs in the ostensible absence of any stimuli that can be shown to elicit or release, or to set the occasion for its occurrence.

- **spontaneous generation**
the production of living matter or organisms from nonliving matter. Prior to the work of Pasteur, it was widely held that spontaneous generation occurs.

- **sporangium**
a capsule in fungi and plants in which meiosis occurs and haploid spores develop.

- **spore**
in the life cycle of a plant or alga undergoing alternation of generations, a meiotically produced haploid cell

that divides mitotically, generating a multicellular individual, the gametophyte, without fusing with another cell.

- **sporophyll**
  a spore-producing leaf specialised for reproduction.

- **sporophyte**
  the multicellular diploid form in organisms undergoing alternation of generations that results from a union of gametes and that meiotically produces haploid spores that grow into the gametophyte generation.

- **spreading centre**
  on plate tectonics, a linear, principally submarine ridge system at which new oceanic crust is added and from which adjacent tectonic plates diverge as a result of convective upwelling of magma.

- **stability**
  often used to mean constancy; more correctly used to mean the propensity to return to a condition (a stable equilibrium) after displacement from that condition.

- **stabilising selection**
  1. the elimination of phenotypes of either extremes from the population by natural selection and hence of the genotypes that produce them; extreme deviants might include the largest as well as the smallest specimens, longest and shortest beaked and so forth.
  2. natural selection that favours intermediate variants by acting against extreme phenotypes.

- **stable equilibrium**
  in an ecological context, a level of a population, or populations, or of resources, which is returned to after slight displacements from that level.

- **stable limit cycles**
  a regular fluctuation in abundance, the path of which is returned to after slight displacements from the path.

- **stakeholders**
  groups, individuals or organisations who may be affected by a development proposal, whether or not their stake in the outcome is explicit.

- **stamen**
  the pollen-producing male reproductive organ of a flower, consisting of an anther and filament.

- **stand**
  1. to assume or maintain an upright position on extended legs.
  2. a group of trees that occupies a specific area and is similar in species, age and condition.

- **standard deviation**
  the square root of the variance.

- **standards and guidelines**
  requirements found in a Forest Plan which impose limits on natural resource management activities, generally for environmental protection.

- **standing crop**
  the biomass of living organisms within a unit area.

- **starch**
  an a-linked polymer of glucose; used by plants as a means of storing energy.

- **stasigenesis**
  no anagenesis or cladogenesis in a lineage across evolutionary time.

- **stasipatric speciation**
  a type of sympatric speciation in which isolation is achieved by karyotype reorganisation.

- **stasis**
  little change in a lineage over long periods of time.

- **static life table**
  a life table constructed from the age structure of a population at a single moment in time.

- **statocyst**
  a type of mechanoreceptor that functions in equilibrium in invertebrates through the use of statoliths, which stimulate hair cells in relation to gravity.

- **statolith**
  1. a starch grain in a plant cell that settles to the low point of the cells in relation to gravity.
  2. in animals, a granule of sand or limestone that, in responding to gravity, stimulates hair cells of a statocyst.

- **status badge**
  see **badge of status**.

- **status signal**

  a term applied to a sign that displays a specific socially relevant characteristic to conspecifics, such as position within a rank order or age or experience of a prospective mate. Thus the highest-ranking male in a group of gorillas displays a conspicuous silver-grey marking on its back (silverback male), in contrast to the black coloration of the rest of the members. In many songbirds there is evidence that colour markings on the feathers signify rank position within a group. In other songbirds the older males that have more breeding experience possess larger song repertoires and in competition with other males acquire larger or better-quality territories.

- **stele**

  the central vascular cylinder in roots where xylem and phloem are located.

- **stellate ganglion**

  a group of cells found in the nervous system of molluscs such as squid that controls the escape response. The stellate ganglion has projections to the muscles of the mantle that allow rapid contraction for quick escape.

- **stenotopic**

  of a species or population, restricted to or capable of persistence in a narrow range of conditions or habitats; see **Eurytopic.**

- **steppe**

  treeless plains of south-eastern Europe and Siberia.

- **stereoisomer**

  a molecule that is a mirror image of another molecule with the same molecular formula.

- **stereotypy**

  1. a term applied when members of a set of successive instances of a response do not vary in their quantitative topographic characteristics.

  2. a behaviour is usually considered stereotyped if the coefficient of variation is less than 30%.

- **steroids**

  a class of lipids characterised by a carbon skeleton consisting of four rings with various functional groups attached.

- **stewardship**
caring for the land and its resources to pass healthy ecosystems to future generations.

- **stigma**
the sticky tip of the carpel that receives the pollen in a flower.

- **stimulus**
any form of stimulus energy that can be detected by the body, requires sensory transduction.

- **stimulus filtering**
selection, from the total stimulation presented by a situation of those stimuli that elicit and regulate behaviour (See **releasing stimulus; sign stimulus**). The filtering can occur in the sense organ, where the receptors exclude part of what they receive or block transmission of its effects (peripheral filtering). But as a rule at least some of the filtering takes place in the central nervous system, although in most cases little is known about where and how such central filtering occurs. It probably takes place in several successive steps. The capacity of nerve cells and neural networks not to respond to some stimuli.

- **stimulus generalisation**
1. a graded series of stimuli of decreasing effectiveness in eliciting a response. For example, an animal is trained to respond to a tone of 1000 Hz but will also respond to lower and higher frequencies without prior training.
2. the behavioural fact that a response conditioned to one stimulus (or set of stimuli) will be elicited by or will occur in the presence of another stimulus (or set of stimuli) which is similar to the conditioned stimulus or discriminative stimulus although there has been no specific training to it.

- **stimulus habituation**
eventually following repeated presentations of a stimulus the given response wanes. This stimulus-specific waning occurs in the sensory organ and the primary sensory centres; when the stimulus is withdrawn the sensitivity returns.

- **stimulus specificity**
stimulus selectivity. The selective responsiveness to stimulation characteristic of a receptor or a behaviour pattern. The tuning characteristics of physiological or behavioural response systems.

- **stimulus summation**
in ethology this usually refers to the additive effects of sign stimuli arriving together or in quick succession, as reflected in the strength or kind of response they elicit.

- **stimulus threshold**
the class of those values of quantified stimuli that will elicit some defined constant response at a fixed strength of less than maximal value.

- **stimulus**
1. a reversible change in the external or internal environment which is above a certain threshold and elicits one or more sensory excitations.
2. a physical event impinging on the receptors of an animal and capable of exciting those receptors.
3. a specified part, or change in a part, of the environment correlated in an orderly manner with the occurrence of a specified response.
4. loosely used as synonymous with stimulus object (an object which produces stimuli) and with stimulus event (an event which produces stimuli).

- **stimulus-response correlation**
an observed relationship between a stimulus and a response, such that a particular response can be shown to be dependent for its occurrence upon the just previous or concomitant occurrence of a specific stimulus or class of stimuli and to vary with variations in the conditions of presentation.

- **stochastic analysis**
statistical computation of the transition probabilities between items in mutually exclusive and collectively exhaustive sets, such as letters of the alphabet or the preening movements of a bird or fly. In applying such analysis to behavioural se-

quences, certain requirements have to be met such as that of stationarity, meaning that the transition probabilities must be stable, at least for the time covered by the sample of data. When the frequency of one item following another is greater or less than chance, the dependency is said to conform to a mark off chain model. If such a different-from-chance transition probability between two items is independent of the preceding items, the dependence is said to be a first-order mark off chain; if the probability of B given A differs according to what immediately precedes A, the dependence is that of a second-order mark off chain; and dependence extending farther back is labelled according to the same pattern.

- **stochastic forces**
  random processes that affect community structure.
- **stochastic process**
  a process where the various outcomes can be predicted with a specified probability, but where the specific event cannot be predicted exactly (deterministically.)
- **stocking level**
  the number of tree in an area as compared to the desirable number of trees for best results, such as maximum wood production.
- **stoloniferous**
  bearing stolons.
- **stolons**
  horizontally growing, short-lived, stems that root at the nodes. Usually at or on the surface of the soil.
- **stoma**
  1. a microscopic pore surrounded by guard cells in the epidermis of leaves and stems that allows gas exchange between the environment and the interior of the plant.
  2. pore in the epidermis of plants through which gas exchange takes places; especially abundant on leaves.
- **stotting**
  a behaviour which occurs in artiodactylids especially during fright, sudden disturbances and play. It consists of jumping from 2 to 12 times

straight into the air with all four legs held stiff, the neck erect and the tail held in a horizontal to vertical position. The function of the behaviour is not known. The sound of the hoofs striking the ground may serve to alert others or the visual exposure of the bright rump patches are possible explanations. It is also possible that the jumping allows the animal to monitor the source of disturbance in an area where the visibility is poor. The behaviour is seen in open grassland species. Alternatively the behaviour may demonstrate condition to the predator and prevent pursuit.

- **strain**

  a selectively inbred stock.

- **straining**

  a method of feeding in geese, some ducks and flamingos in which small organisms are passed through a sieve-like structure on the beak.

- **strategy**

  in behavioural science this term has acquired a specialised meaning.

1. formerly it was used only in roughly the same sense as 'way of life'. One still speaks of foraging strategies or reproductive strategies (monogamy, polygamy, family).
2. but in socio-biology the term has become associated with mathematical game theory. Roughly speaking, 'strategy' in this sense refers to a comprehensive behavioural plan to deal with decision making; for each decision situation it prescribes a rule or an action or at least specifies the probabilities of results of different actions for the different situations. The game theory of evolutionary biology poses the question of how far strategies for dealing with conflicts between two or more individuals can be understood as adaptive. It is also concerned with how natural selection 'evaluates' strategies. Complicating an analysis of this sort is the fact that not only are strategies adapted to the world (nonliving and living) outside the species, they also concern situations between members of a species; in many cases

the decision is affected by what others are doing. To bring calculation to bear on this problem, the concept of Evolutionarily Stable Strategy (ESS) has been developed. In the simplest case conceivable (assuming that most of the individuals of a population will adopt the strategy), one strategy is superior to all others. In this case an alternative strategy has no chance of becoming established or of displacing the first, which is therefore an ESS. A genetically distinctive set of rules for behaviour exhibited by individuals.

- **stratigraphic sequence**
  over time sediments are deposited in layers which make up a stratigraphic section. A fossil type found in each layer of deposits of a stratigraphic section becomes a stratigraphic sequence and can be studied for changes over time.

- **stratigraphy**
  the arrangement of rock strata especially with regard to the chronological order of the sequence.

- **stress**
  1. physics has a strict definition of 'a force per unit area' and producing 'strain' in the body to which the force is applied. Biology has a wide variety of meanings, e.g. any condition that results in reduced growth, any condition that prevents an organism from realizing its 'genetic potential'. The word is often redundant, e.g. the effects of drought stress/the effects of drought. The word is also often used confusingly in two senses, both to describe force and the condition induced in the organism by the force - a confusion of stimulus and response. We have tried not to use the word.
  2. a physiological condition that results from excessive environmental or psychological pressures or disturbances. This may be a pathological response or a specific and evolved adaptation for coping with prolonged life threatening conditions.

- **stretching**
  behavioural sequences associated with extending the limbs or body musculature.

- **strict**
1. (aerobe) an organism that can survive only in an atmosphere of oxygen, which is used in aerobic respiration.
2. (anaerobe) an organism that cannot survive in an atmosphere of oxygen; usually uses sulphate or nitrate, as the terminal electron acceptors in the generation of ATP.

- **stridulation**
sound production by rasp and scraper method most often in insects. Rubbing one body surface against another.

- **stringer**
a strip of vegetation different from surrounding vegetation, such as a stringer of aspen in a area of spruce.

- **stroma**
the fluid of the chloroplast surrounding the thylakoid membrane; involved in the synthesis of organic molecules from carbon dioxide and water.

- **stromatolite**
1. a rock-like formation formed by communities of cyanobacteria and other prokaryotes in shallow water; formed during the Precambrian era.
2. rock made of banded domes of sediment in which are found remnants of the most ancient forms of life: prokaryotes dating back as far as 3.5 billion years.

- **structural**
1. a type of molecular notation in which the constituent atoms are joined by lines representing covalent bonds.
2. a gene that codes for a polypeptide.

- **structural diversity**
range of types of physical structure in a community that may provide habitats for species.

- **structural gene**
1. the DNA code unit whose primary product is messenger RNA and which thus controls synthesis of a particular protein or polypeptide.
2. a gene that determines the primary structure (i.e. the amino acid sequences) of a polypeptide. See **operon, regulatory gene**.

- **structure**
  how the parts of ecosystems are arranged, both horizontally and vertically. Structure might reveal a pattern, or mosaic, or total randomness of vegetation.

- **structured demes**
  individuals interact with only a very small subset of a total deme. The deme then is broken up into many smaller trait groups in which actual ecological interactions occur.

- **style**
  the slender, neck like portion of the carpel that leads to the ovary.

- **stylet**
  slender, elongated mouth part, usually of an insect, used for stinging or piercing prey or sucking sap.

- **subatomic particle**
  any of a variety of the fundamental units of matter and energy associated with atoms. Includes protons, neutrons and electrons.

- **subduction zone**
  in plate tectonics an elongate region where the lithosphere plate sinks below another part of the plate; where the plate descends into the mantle under the leading edge of another plate.

- **subjective probability**
  the subjects evaluation of the consequences of an action that it might take.

- **submissive behaviour**
  behaviour of subordination in order to prevent attack. Defensiveness.

- **submissive coloration**
  a change of coloration in subordinate individuals.

- **subordinate behaviour**
  the status of an individual in a group does not have access to resources or females. See **dominance**.

- **sub social**
  1. refers to an association, either permanent or temporary, between a mother and her adult offspring.
  2. the sub social route to sociality refers to a long-term cooperative association between mother and offspring.

- **subsong**
  1. fledgling vocalisations which lack the structure of the full song of the adult.

2. low intensity song prior to full song production of adult animals.

- **subspecies**
a set of populations of a species that occupy different geographic areas and possess somewhat different characters.

- **substitute goods**
goods that you might purchase instead of a particular good. For example, different types of bread are substitutes for each other.

- **substitution**
the complete replacement of one allele by another within a population or species; in the term nucleotide substitution, the complete replacement of one nucleotide pair by another within a lineage over evolutionary time. See **fixation**.

- **substrate**
1. the substance on which an enzyme works.
2. (-level phosphorylation) the formation of ATP by directly transferring a phosphate group to ADP from an intermediate substrate in catabolism.

- **successful dispersal**
dispersal is successful if propagule obtains opportunity to breed and raise young (i.e. genes migrate).

- **succession**
1. the non-seasonal, directional and continuous pattern of colonization and extinction on a site by populations.
2. the natural replacement of one plant community with another. Conditions of the prior plant community (or successional stage) create conditions that are favourable for the establishment of the next stage.
3. transition in the species composition of a biological community, often following ecological disturbance of the community; the establishment of a biological community in an area virtually barren of life.
4. (primary) a type of ecological succession which occurs in an area where there were originally no organisms.
5. (secondary) a type of ecological succession that occurs where an existing community has been severely

cleared by some disturbance.

- **successional stage**
  a stage of development of a plant community as it moves from bare ground to climax. The grass-forb stage of succession precedes the woody shrub stage.

- **succulents**
  plants with fleshy or juicy tissues with high water content characteristic of desert and saline environments.

- **suckle**
  the behaviour of young mammals when feeding on milk from the mother.

- **suitability**
  the appropriateness of certain resource management to an area of land. Suitability can be determined by environmental and economic analysis of management practices.

- **sulphur bacteria**
  bacteria which live in environments where oxygen is scarce or absent and which metabolise compounds containing sulphur. Includes purple and colourless forms.

- **sulphur cycle**
  cyclic movement of sulphur in different chemical forms, from the environment to organisms and then back to the environment.

- **summation**
  a phenomenon of neural integration in which the membrane potential of the postsynaptic cell in a chemical synapse is determined by the total activity of all excitatory and inhibitory presynaptic impulses acting on it at any one time.

- **sun bathing or sunning**
  behaviour that is associated with animals seeking out and basking in sunlight. Associated with thermoregulation.

- **supergene**
  a group of two or more loci between which recombination is so reduced that they are usually inherited together as a single entity.

- **super-normal sign stimulus**
  a term applied to certain sign stimuli that have proven amenable to quantification along some scale. The sign stimulus, as it occurs in the field, falls at

some point on this scale. To stimuli below this value, strength or intensity of response is less. If responses are given at greater strength or intensity to stimuli above this value, the stimuli of these magnitudes are referred to as 'supernormal stimuli' (that is, they are more effective than 'normal' stimuli).

- **supernormal stimulus**
  a sign stimulus that is more effective in eliciting a response than naturally occurring actions or objects should be. A stimulus that elicits a particular behaviour pattern more easily than the natural releaser which normally elicits the behaviour. The innate releasing mechanism responds specifically to the characteristics of the object or situation for which the reaction has evolved.
- **supernova**
  a stellar explosion.
- **supernumerary**
  a member of a social unit in addition to the breeding pair.
- **superorganism**
  any colony possessing features of social organization analogous to the physiological properties of a single organism; any group of organisms acting as a single functional unit.
- **superorganism concept**
  the view of communities as consisting of member species that are tightly bound together both now and in their common evolutionary history.
- **superspecies**
  a group of semispecies.
- **superstitious behaviour**
  1. unless an experimenter is very careful, during shaping or approximation conditioning of a rat, or a pigeon, or a human subject, he may reinforce a response in which he is not interested or reinforce too often one of the responses that is in the approximation sequence of responses. These responses, occurring henceforth at a relatively high rate or in great strength, are referred to as superstitions or as superstitious responses.
  2. if stimuli that are usually reinforcing (e.g., food) are randomly delivered to a pi-

geon over a long period of time irrespective of his behaviour, at the end of the period the subject can be observed to repeat over and over some response. Such a response is termed superstition.

- **'supertramp species'**
species with good colonising ability that arrive earliest in a new habitat.

- **supply curve**
the graphical representation of the supply function. Because producers would like to sell more at higher prices, the supply function slopes upward.

- **supply function**
the mathematical function that relates price and quantity supplied for goods or services. The supply function tells how many units of a good that producers are willing to produce and sell at a given price.

- **suppressor**
(T cell- TS) a type of T cell that causes B cells as well as other cells to ignore antigens.

- **suppressor gene**
a gene that can reverse the effect of a specific type of mutation in other genes.

- **suppressor mutation**
a mutation that totally or partially restores a function lost by a primary mutation and is located at a genetic site different from the primary mutation.

- **suprachiasmatic nucleus**
part of the hypothalamus which is linked by neurons to cells of the eye and plays a role in tracking the length of the day and regulating the sleep-wake cycle.

- **surface resources**
renewable resources that are on the surface of the earth, such as timber and forage, in contrast to ground water and minerals which are located beneath the surface.

- **surface tension**
a tautness of the surface of a liquid, caused by the cohesion of the molecules of liquid. Water has an ex-

tremely high surface tension.

- **surplus yield model**
  a simple model of the impact of harvesting on a population, in which the population is represented by its size or biomass, undifferentiated into any internal structure.

- **surrogate resource**
  a resource, not in itself in limited supply, which is competed for because of the access it provides to some other resource, which is or may become limited in supply.

- **survival value**
  the survival of a trait within a population depends upon the extent to which the trait contributes to reproductive success.

- **survivorship**
  the probability of a representative newly born individual surviving to various ages.

- **survivorship curve**
  1. a plot of the declining size of a cohort, or presumed cohort, as the individuals die, usually with time on the horizontal axis and $\log_{10}$ l_x on the vertical axis (where l_x is the proportion of the original cohort still alive).
  2. a plot of the number of members of a cohort that are still alive at each age; one way to represent age-specific mortality.

- **suspension-feeder**
  an aquatic animal, such as a clam or a baleen whale, that sifts small food particles from the water; also called filter-feeder.

- **sustainability**
  1. the ability of an ecosystem to maintain ecological processes and functions, biological diversity and productivity over time.
  2. ability of a system to survive for some specified (finite) time.

- **sustainable**
  1. referring to an activity that is able to be carried out without damaging the long-term health and integrity of natural and cultural environments.
  2. the yield of a natural resource that can be produced continually at a given intensity of management is said to be sustainable.

3. meeting the needs of the present without diminishing the ability of people, other species or future generations to survive.

- **sustainable agriculture**

1. long-term productive farming methods that are environmentally safe.
2. method of growing crops and raising livestock based on organic fertilisers, soil conservation, water conservation, biological control of pests and minimal use of non-renewable fossil-fuel energy.

- **sustainable development**

1. this term is often used interchangeably with ecologically sustainable development, but has a more human-oriented focus. It has been defined as development that meets the needs of the present generation without compromising the ability of future generations to meet their own needs.
2. forms of economic development and activities that do not deplete or degrade the natural resources upon which present and future economic growth and life depend.

- **sustainable living**

taking no more potentially renewable resources from the natural world than can be replenished naturally and not overloading the capacity of the environment to cleanse and renew itself by natural processes.

- **sustainable society**

a society that manages its economy and population size without doing irreparable environmental harm by overloading the planet's ability to absorb environmental insults, replenish its resources and sustain human and other forms of life over a specified period-usually hundreds to thousands of years. During this period it satisfies the needs of its people without depleting earth capital and thereby jeopardising the prospects of current and future generations of humans and other species.

- **sustainable system**

a system that survives and functions over some specified (finite) time; a system that attains its full expected lifetime.

- **sustainable yield (sustained yield)**

1. highest rate at which a potentially renewable resource can be used without reducing its available supply throughout the world or in a particular area.
2. the yield that a renewable resource can produce continuously at a given intensity of management.

- **swarming**

process of colony fission in honey bees. Also refers to a mating system of flies in which individuals (usually males) fly in close proximity usually over a high point in the environment with the effect of attracting females.

- **switching**

of a predator, the tendency to switch between prey categories according to their relative abundance in the environment.

- **switching site**

break points at which gene segments combine in gene rearrangements.

- **symbiont**

the smaller participant in a symbiotic relationship, living in or on the host. See **symbiosis**.

- **symbiosis**

1. a close association between the individuals of pairs of species. The term 'mutualism' is reserved for symbioses for which there is evidence that the association brings mutual gains.
2. an association between two or more organisms which benefits at least one of them.
3. a general term referring to any association among organisms of different species whether parasitic or mutualistic.
4. an ecological relationship between organisms of two different species that live together in direct contact.
5. literally means 'living together' in Latin. Any intimate relationship or association between members of two or more species. The members of the relationship are symbionts. Obligate symbionts rely so heavily on the relationship that they cannot feed, reproduce, or perform some other crucial life function in the absence of their

symbiotic partner(s). There are three main categories of symbiosis: *commensalism*, *mutualism* and *parasitism*, with some degree of blending at the edges of these definitions in many cases.

- **symbiotic**

refers to a component or member of a system of symbiosis. These organisms have a symbiotic relationship.

- **symbolic behaviour**

1. verbal behaviour.
2. a hypothesised class of behaviour that cannot necessarily be directly observed.
3. the observed behaviour from which symbolic behaviour is inferred.
4. loosely and colloquially, thinking.

- **sympathetic nervous system**

one of two divisions of the autonomic nervous system of vertebrates; generally increases energy expenditure and prepares the body for action.

- **sympatric speciation**

1. development of a genetic discontinuity within a population even in the absence of geographic separation; also See **speciation.**
2. a mode of speciation occurring as a result of a radical change in the genome that produces a reproductively isolated subpopulation in the midst of its parent population.
3. speciation without geographic isolation; the acquisition of isolating mechanisms within a deme.

- **sympatry**

1. the presence of two or more species living in such proximity that breeding between them should be possible, though their continued existence as separate species indicates that it does not normally happen. This contrasts with allopatry in which regional or geographical isolation normally denies the possibility of interbreeding.
2. pertaining to two or more populations occurring in the same geographical area (allopatric, parapatric).

- **symplast**

in plants, the continuum of cytoplasm connected by plasmodesmata between cells.

- **symplesiomorphy**
  shared ancestral character state (from cladistics.)

- **synapomorphy**
  a shared, derived character (from cladistics.)

- **synapse**
  1. communicating junction between two nerve cells, between receptor and nerve or between nerve cell and effector organ.
  2. Electrical synapses between two cells have an immediate, physical connection whereas chemical synapses have a synaptic cleft over which chemical transmitters travel from the presynaptic to the postsynaptic side.
  3. specialised site at which communication between one neuron and another takes place.
  4. the locus where one neuron communicates with another in a neural pathway; a narrow gap between a synaptic terminal of an axon and a signal-receiving portion of another neuron or effector cell. Neurotransmitters released by synaptic terminals diffuse across it, relaying messages to the postsynaptic cell.

- **synapsid**
  a predatory reptile that evolved during the Permian period and gave rise to several lineages, including the mammal-like therapsids.

- **synapsis**
  the pairing of replicated homologous chromosomes during prophase I of Meiosis.

- **synaptic**
  1. (terminal) one of many swollen bulbs at the end of an axon, in which neurotransmitter is stored and released.
  2. (vesicle) one of many membrane-enclosed sacs in a synaptic terminal that contains neurotransmitter.

- **synaptic plasticity**
  changes in excitability and transmetre release at synaptic junctions between neurons.

- **synaptomenal complex**
  a protein structure, visible in electron micrographs of prophase I of meiosis; function is unknown but correlated with, and essential for,

chiasmata formation and crossing over; often zipper like in appearance.

- **synchronisation**
  temporal coordination.

- **synergism**
  1. the situation in which the combined effect of two forces, e.g. treatment with two drugs, is greater than the sum of their separate effects.
  2. combined action of two where one does not work alone.

- **syngamy**
  the process of cellular union during fertilisation.

- **syntactics**
  in linguistics, the rules governing how words are combined from phrases and sentences. In communication the study of relations among signals. In bird song, how the component parts of songs are organized. In communication, the study of relations among signals.

- **syrinx**
  sound-producing organ in birds located where the trachea joins the bronchi of the two lungs.

- **systematics**
  1. the comparative study of the historical, evolutionary and genetic relationships among organisms and of their phenotypic similarities and differences.
  2. the branch of biology that studies the diversity of life. Systematics encompasses taxonomy and is involved in reconstructing phylogenetic history.

- **systemic mutation**
  a mutation, postulated by R. Goldschmidt, that would fundamentally reorganise the germ plasm and permit the origin of wholly new types of organisms.

- **systole**
  the stage of the heart cycle in which the heart muscle contracts and the chambers pump blood.

- **t**
  amount of time elapsed (may be min, hr, d, yr, generation.)

- **tabula rasa**
  the notion that the brain at the time of birth is without pre-programmed

behavio-ural elements and that all behaviour is the result of information acquired during the individual's life.

- **tactic**
one of the options that may be exercised by an animal whose behaviour is guided by a conditional strategy.

- **taiga**
1. the coniferous forest that extends across much of North America and Eurasia bounded by tundra to the North and by steppe to the south.
2. a type of coniferous evergreen forest, found in the sub arctic areas of North America and Eurasia. Also known as a boreal forest.
3. the coniferous or boreal forest biome, characterised by considerable snow, harsh winters, short summers and evergreen trees.

- **taming**
a process that progressively reduces and eliminates animal's initial tendencies to flee and act negatively toward humans.

- **tandem running**
behaviour found in social insects in which one individual follows closely behind another during exploration or recruitment.

- **tannins**
complex astringent substances containing phenolic compounds found in plants and usually making the plant material less readily digested. Especially common in bark, unripe fruits and galls on woody plants.

- **target**
a National Forest's annual goals for accomplishment for natural resource programs. Targets represent the commitment the Forest Service has with Congress to accomplish the work Congress has funded, and are often used as a measure of the agency's performance.

- **target cell**
a cell bearing specific receptors for a particular signal, such as a hormone.

- **taste**
contact chemoreception.

- **taste aversion**

a form of conditioned learning, bait shyness. Learned avoidance of a particular type of food that has caused an animal to become ill after eating it.

- **taxis**

1. act of orienting towards some external stimulus or combination of stimuli. Spatial orientation, aided by different sensory modalities, is described by the corresponding term e.g. relative to light (phototaxis), smell (chemotaxis), sound (phonotaxis), or gravity (geotaxis). If orientation is towards the source, it is called a positive taxis, and away from the source a negative taxis. In such instances individuals move in a directed fashion along a particular stimulus gradient until they reach a perceived optimal range.
2. a movement toward or away from a stimulus.
3. orientation reaction or movement in freely moving organisms as opposed to tropism in sessile organisms.
4. a term applied to a broad class of behaviours specifiable in terms of the responses (locomotory and orientative) and of the stimuli (most often visual) controlling them.

- **taxon**

1. any group of organisms to which any rank of taxonomic name is applied.
2. the named taxonomic unit to which individuals or sets of species are assigned; may be a species, genus, family or whatever. Higher taxa are those above the species level.

- **taxon cycle**

replacement of species of a given taxon within a community over evolutionary time, such that predominant species vary cyclically in character.

- **taxonomy**

1. the study of the rules, principles and practice of classifying living organisms.
2. the naming and assignment of organisms to taxa.
3. the branch of biology concerned with naming and classifying the diverse forms of life (the practitioners are called taxonomists).
4. (classical evolutionary taxonomy) An approach to

biological classification and phylogeny that considers both overall homology and evolutionary branching sequences.
5. the classification of living organisms according to the hierarchy of relationships.

- **tectonic**
refers to the movement of the plates that comprise the earth's crust and to the deformation of those plates.

- **tectonic plate**
an area of the earth's crust which moves during geological time resulting in continental drift and other major changes in the topography of the surface of the globe.

- **teleology**
the doctrine that natural phenomena result from or are shaped by design or purpose.

- **teleonomy**
the scientific investigation of programmed purposiveness in living organisms and in certain man-made machines. Behaviour patterns are quite prominently teleonomic; that is, they often appear 'goal-directed' yet can be fully accounted for in terms of antecedent proximate causation. Teleonomic behaviour is governed either by a genetically determined (closed) or experientially acquired (open) program, and through this program behaviour is adjusted in fine detail to the prevailing environmental conditions to arrive at some end state; in this sense the behaviour is purposive. Such adjustment is often achieved through the guidance furnished by feedback.

- **telotaxis**
a type of orientation in which the animal makes a direct approach to the stimulus source. The process by which animals attain directional orientation without having to compare the stimulation from two receptors. It is found in animals that have sensory organs such as the vertebrate eye that are capable of providing directional information as a result of their structure. (DM).

- **temperate**
1. (deciduous forest) a biome located throughout midlati-

tude regions where there is sufficient moisture to support the growth of large, broad-leaved deciduous trees.

2. (virus) a virus that is reproduced without killing the host cell.

- **temperate rain forest**

  a type of forest found in only a few places around the world, such as the Pacific temperate rain forest on the west coast of North America. These forests are often dominated by conifer trees adapted to wet climates and cool temperatures.

- **temperature**

  a measure of the intensity of heat in degrees, reflecting the average kinetic energy of the molecules.

- **template**

  a term used in bioacoustics for information placed in memory that provides a model against which an animal can match its own production during vocal development. In many songbirds the young males have to learn their songs, often quite early in life, from the father or other exemplars. This learning often occurs far in advance of initial song production, in many cases during a limited sensitive period (See **Song Imprinting**). At this time the young animal acquires a template, or 'sound print' of the exemplar by ear, to which it compares its auditory feedback in shaping its own song during later development. Artificially deafened songbirds which cannot hear their own song and cannot compare it to the template, develop very atypical songs as a rule. For bird species in which the song is at least partly genetically programmed, some authors talk of an innate template that maybe completed or modified through experience.

- **temporal patterning**

  the structure of behaviour based on its timing characteristics, usually inter-act interval.

- **temporal summation**

  1. in neurons, summation of two or more inputs of the same type that occur at different times.

2. an accumulative effect on the membrane potential of a postsynaptic cell; caused by chemical transmissions from the synaptic terminal(s) occurring so close together in time that each postsynaptic potential affects the membrane before the voltage has returned to the resting potential after the previous stimulation.

- **temporal variation**
  variability in conditions, such as temperature, on an hourly, daily or seasonal basis.

- **tendon**
  a type of fibrous connective tissue that attaches muscle to bone.

- **teneral**
  refers to immatures in some groups of insects.

- **teratology**
  the study of abnormal development.

- **terminal bud**
  embryonc tissue at the tip of a shoot, consisting of developing leaves and a compact series of nodes and internodes.

- **termination codon**
  a codon that specifies the termination of translation.

- **termination sequence**
  a DNA sequence at the end of a transcriptional unit that signals the end of transcription.

- **termiticolous**
  inhabiting termite nests.

- **terpenoids**
  aromatic hydrocarbons formed by many plants and responsible for many of the plant scents.

- **terricolous**
  living on or in soil.

- **territorial advertisement**
  refers to all the mechanisms by which individuals defend territories from intrusion by conspecifics including territorial scent marking, singing, and visual displays.

- **territorial marking**
  the posting and designation of a territory with specific chemicals.

- **territorial song**
  singing that advertises possession of a territory to rivals of the same sex, a form of marking behaviour. Ter-

ritorial marking is probably the main function of song in most songbirds and many other animal groups. But much territorial singing also serves other functions as well, such as attracting unmated partners.

- **territoriality**
  the establishment by an animal or animals of an area from which other individuals are partially or totally excluded.

- **territory**
  1. a part of the home range which is defended from intrusion by conspecifics or a specific location in which an individual, the owner, is dominant. A fixed area from which rival intruders are excluded by some combination of advertisement, threat and attack.
  2. any defended area; area of more or less fixed boundaries from which rival conspecifics are excluded.

- **tertiary**
  1. a geological era lasting from approximately 65 to 2 million years ago.
  2. a geological period of the Cenozoic era, 65-16 my ago.

- **tertiary structure**
  irregular contortions of a protein molecule due to interactions of side chains involved in hydrophobic interactions, ionic bonds, hydrogen bonds and disulphide bridges.

- **testcross**
  breeding of an organism of unknown genotype with a homozygous recessive individual to determine the unknown genotype. The ratio of phenoypes in the offspring determines the unknown genotype.

- **testis (pl. testes)**
  the male reproductive organ or gonad, in which sperm and reproductive hormones are produced.

- **testosterone**
  the most abundant androgen in the male body.

- **tetanus**
  the maximal, sustained contraction of a skeletal muscle, caused by a very fast frequency of action potentials elicited by continual stimulation.

- **tethys sea**
the oceans surrounding the super-continent of Pangaea.

- **tetrapod**
vertebrate possessing two pairs of limbs, such as an amphibian, reptile, bird and mammal.

- **thalamus**
one of two integrating centres of the vertebrate forebrain. Neurons with cell bodies in the thalamus relay neural input to specific areas in the cerebral cortex and regulate what information goes to the cerebral cortex.

- **thanotosis**
playing dead; death-feigning behaviour.

- **thecodont**
a reptile that descended from the stem reptiles during the Permian period and that served as the ancestral group of dinosaurs, crocodiles and birds.

- **thelytoky**
all-female species reproduce in this way by parthenogenesis.

- **theory**
1. a set of statements about empirical concepts, relationships among them, and hypotheses postulating further relationships among them. Theories often include statements involving the empirical concepts of a universe of discourse other than that in which the theory is derived.
2. A scientifically accepted general principle supported by a substantial body of evidence offered to provide an explanation of observed facts and as a basis for future discussion.
3. an explanation of observed phenomena which has been accepted as true because significant confirming evidence has been obtained. Examples include the Cell Theory and Theory of Evolution.

- **theory of evolution**
the total body of knowledge relating to phenomena whereby life forms become modified during descent from ancestors.

- **therapsid**
a member of a reptile lineage that descended from

synapsids during the Permian period and eventually gave rise to mammals.

- **thermal cover**

cover used by animals against weather. For elk, thermal cover can be found in a stand of coniferous trees at least 40 feet tall with a crown closure of at least 70%.

- **thermocline**

a narrow vertical zone of rapid temperature change that differentiates the warm upper layer from the cold bottom water of a lake.

- **thermodynamics**

1. the study of energy transformations that occur in matter. First law- conservation of energy: Energy can be transferred and transformed, but not created or destroyed.
2. second law- Energy transformations result in decreases in free energy and increases in entropy.

- **thermoneutral zone**

1. the range of environmental temperatures for an endotherm over which it has to exert the minimum metabolic effort in order to maintain a constant body temperature.
2. the range of environmental temperatures in which an endothermic organism can maintain a stable body temperature without metabolic heat production above that produced in the course of maintenance metabolism.

- **thermoregulation**

maintenance of internal temperature within a tolerable range.

- **thermoregulatory behaviour**

behaviour associated with the animal maintaining its body temperature at an optimal level.

- **thick filament**

a filament composed of staggered arrays of myosin molecules; a component of myofibrils in muscle fibres.

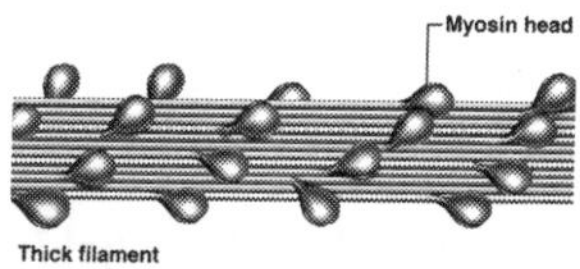

- **thigmotaxis**

a type of orientation in which animals orient by remaining closely associated with environmental features, e.g. a rat running along a wall.

- **thigmotropism**
  the directional growth of a plant in relation to touch.

- **thin filament**
  a filament consisting of two strands of actin and one strand of regulatory protein coiled around one another; one component of myofibrils in muscle fibres.

- **thinning**
  a cutting made in an immature stand of trees to accelerate growth of the remaining trees or to improve the form of the remaining trees.

- **thinning line**
  the line on a plot of log mean individual weight against log density along which self-thinning populations of a sufficiently high biomass tend to progress and beyond which they cannot pass.

- **third generation insecticides**
  synthesised, usually organic insecticides, which are nevertheless produced with the aim of minimising the impact on non-pest, non-target species.

- **thirst**
  a state of motivation which arises primarily as a result of dehydration of the body tissues.

- **Thomas Malthus**
  formulated the concept that population growth proceeds at a geometric rate.

- **threat behaviour**
  1. avoidance behaviour with aggressive motivation, for example intimidation display with elements of threat. Threat behaviour is species typical behaviour.
  2. Those set of behaviours of an animal that have been shown to produce flight behaviour at some strength in another animal (usually of the same species) when they occur in its presence.

- **threat yawning**
  a threat posture occurring in primates and in hippopotamus, in which the mouth is opened widely as in yawning but the behaviour is not released by the physiological factors that release yawning.

- **threatened species**

1. those plant or animal species likely to become endangered throughout all or a specific portion of their range within the foreseeable future.
2. wild species that is still abundant in its natural range but is likely to become endangered because of a decline in numbers.

- **threshold**

1. the strength of a stimulus which just results in a measurable characteristic response.
2. minimum depolarisation of the membrane potential necessary to evoke an action potential in a neuron.
3. the potential an excitable cell membrane must reach for an action potential to be initiated.

- **thrust**

a part of sexual behaviour during mounting where the male moves the pelvis so as to touch the female with the intromittant organ this may occur before, during or after intromission.

- **thwarting**

1. in ethology this term usually refers to a situation in which appetitive behaviour fails to achieve the stimulus conditions necessary for completion of the activated behaviour sequence.
2. more generally, any situation where ongoing behaviour is frustrated in reaching its proximate goal by absence of the requisite stimulation. Its effects are similar to those of conflict among incompatible tendencies in that it can be the occasion for displacement activity, for example.

- **thylakoid**

a flattened membrane sac inside the chloroplast, used to convert light energy to chemical energy.

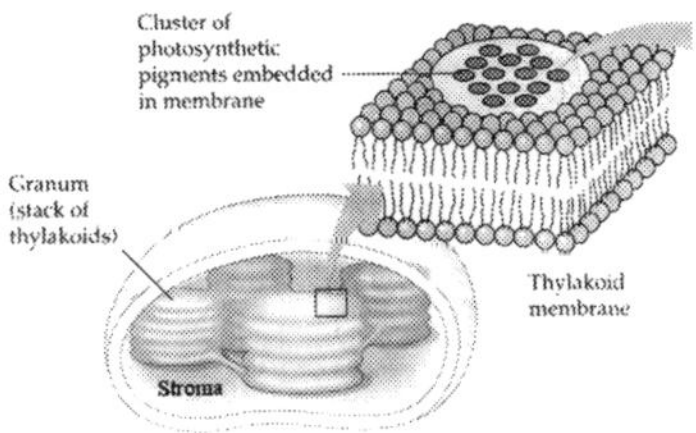

- **thymus**

an endocrine gland in the neck region of mammals that is active in establishing the immune system; secretes several messengers, including thymosin, that stimulate T cells.

- **thyroid-stimulating hormone**
  a hormone produced by the anterior pituitary gland that regulates the release of thyroid hormones.

- **ti plasmid**
  a plasmid of a tumour-inducing bacterium that integrates a segment of its DNA into the host chromosome of a plant; frequently used as a vector for genetic engineering in plants.

- **tidbitting**
  a display by gallinaceous birds, originally consisting of pecking at the ground and calling, used for enticing chicks to approach and feed. In many species the behaviour has been further ritualised and incorporated into courtship patterns, where it serves the secondary function of luring the female to the male.

- **tight junction**
  a junction between epithelial cells, in which there is no gap whatever between the adjacent cells. Materials may get through a tight junction only by entering the epithelial cells themselves.

- **tiller**
  a branch formed at or near ground level by grasses and sedges.

- **timber classification**
  the classification of forested lands into land management alternatives according to how the land relates to management of the timber resource there.

- **time constant**
  time it takes a change in membrane potential to reach 63% of its final value or the time it takes for it to decay to 37% of its peak value. It determines how long a synaptic signal lasts.

- **time sharing**
  a situation in which one of the activities controls when the other can occur. When the causal system for the controlling or dominant activity is in operation, it inhibits that of the other subdominant activity, which can be expressed only during pauses in the dominant activity, when the subdominant activity is disinhibited.

Interruption of activities can indicate which one is dominant in a time-sharing situation. If the interruption occurs during performance of the subdominant activity and does not extend past the point of recurrence of the dominant activity, the temporal organization of the behaviour will not be affected-the dominant behaviour will resume on schedule. If the interruption occurs during the dominant activity, the next pause in that activity will be correspondingly delayed, thus altering the timing of the subdominant activity. Experiments using this approach are called masking experiments; they have produced evidence of time sharing in a variety of contexts, such as courtship in sticklebacks and feeding and drinking in doves.

- **time-delay population model**

  a model of population growth in which the net reproductive rate of individuals currently is determined by the population size some time previously.

- **tissue**

  an integrated group of cells with a common structure and function.

- **tit-for-tat**

  This strategy is simply one of cooperating on the first move and then doing whatever the other player did on the preceding move. Thus tit for tat is a strategy of cooperation based on reciprocity.

- **t-maze**

  a T-shaped pathway basically similar to a runway. The starting box is at the base of the T and goal boxes are at each end of the cross-piece. On a trial, food or water may be placed in either or both goal boxes. Discriminative stimuli are sometimes placed in the arms of the T.

- **tobogganing**

  a locomotory pattern of penguins in which they slide on their stomach, propelling themselves forward with their feet.

- **tolerance (successional)**

  where modification of the environment by early occupants has little or no effect on subsequent performance of late-successional species.

- **tongue-flicking**
  in many snakes tongue-flicking serves an orientation and olfactory function and involves the use of the Jacobson's organ.

- **tonic**
  refers to a neuron in which action potentials occur again and again.

- **tonic immobility**
  temporary inability to move manifested by an animal upon close exposure to or handling by some powerful agent such as a predator; death feigning.

- **tonic receptor**
  type of receptor cell in which adaptation is slow that the response of the cell to the stimulus does not declines quickly. See **phasic receptor.**

- **tonotopic**
  in the auditory system, the hearing organ is mapped point-to-point onto the auditory portion of the CNS.

- **tool use**
  the fashioning of an object external to the body of the animal to accomplish some end. The manipulation of objects for purposes such as care of the body or acquisition and preparation of food. Tool use has been described for a large number of animal species and in connection with many different kinds of behaviour. Hermit crabs slip into empty gastropod shells, which they carry around as a means of protection for their vulnerable abdominal segments. The Egyptian vulture throws stones at ostrich eggs to smash them and get at their edible contents; elephants hold sticks with their trunks to scratch their backs; and chimpanzees have been seen to use sticks to strike at leopards or pile boxes on top of one another to reach food hanging out of reach. The best-known cases of tool use include that of the Galapagos woodpecker finch, which takes a cactus spine in its bill and pokes it into bark to get insects, and the North American sea otter, which breaks open shellfish by holding them in its forepaws and dashing them against a stone

cradled on its chest as it swims on its back. Still further examples have been reported for wasps, blue jays, and baboons. Sometimes even conspecifics are used as tools, weaver ants use their mandibles to pick up larvae, whose silk glands they apply to the edges of rolled-up tobacco leaves, working the larvae back and forth like weaving shuttles to sew the edges of leaves together. As adults the ants no longer possess silk glands. Some animals go so far as to fashion or improve the tools they use. Thus chimpanzees break off twigs from which they peel the bark and remove side branches before poking them down holes in termite nests to fish for termites. Also the woodpecker finch can break off the cactus spine it uses as a probe.

Cases have been reported of animals using objects as tools in novel ways apparently arrived at by perception of configural relationships rather than trial and error. Such cases have been cited as instances of insight learning. There are also cases of a particular kind of tool use being handed on from one generation to the next through tradition.

- **topographic map**
  the point-to-point representation of a sensory surface in the central nervous system.

- **topography**
  representation of the physical structure of an environment.

- **topography (of a response)**
  the full quantitative specification of all the relevant, physically measurable dimensions of a response.

- **total catchment management**
  an integrated management approach which aims to coordinate all activities and processes occurring in a water catchment. These processes and activities include use of the land, water and related vegetation resources. Coordination occurs on a catchment basis so as to balance resource use and nature conservation objectives. It provides the framework for the community, industry and govern-

ment to work together to overcome environmental and resource management problems.

- **total economic value**
  the sum of all types of use and non-use values for a good or service.

- **Total Fertility Rate (TFR)**
  estimate of the average number of children that will be born alive to a woman during her lifetime if she passes through all her childbearing years (ages 15-44) conforming to age-specific fertility rates of a given year. In simpler terms, it is an estimate of the average number of children a woman will have during her childbearing years.

- **totipotent**
  referring to embryonic cells that retain the potential to form all parts of the animal.

- **trace element**
  an element indispensable for life but required in extremely minute amounts.

- **trace fossils**
  these are fossils known only from their traces or the tracks they leave in the sediments such as footprints, burrows and paths rather than from the physical impression of the organism or its hard part remains.

- **trachael system**
  a gas exchange system of branched, chitin-lined tubes that infiltrate the body and carry oxygen directly to cells in insects.

- **trachea**
  1. the windpipe; that portion of the respiratory tube that has C-shaped cartilaginous rings and passes from the larynx to two bronchi.
  2. tiny air tubes that ramify throughout the insect body for gas exchange.

- **tracheid**
  a water-conducting and supportive element of xylem composed of long, thin cells with tapered ends and walls hardened with lignin.

- **tractor logging**
  a logging method that uses tractors to carry or drag logs from the stump to a collection point.

- **tradeoffs**
  the notion that animals cannot be equally good at everything, i.e. they might re-

duce disease by staying warm but they increase metabolism and thus reduce the amount of energy available for other activities such as reproduction.

- **tradition**
  knowledge learned by an individual is retained among the group members beyond the life span of the individual who introduced it. This transfer of knowledge or information from one individual to another and one generation to the next is referred to in ethology as tradition.

- **trail pheromone**
  a chemical substance laid down as a trail by one individual and followed by another member of the same species.

- **train**
  to subject an animal to experimental procedures such that one or more of its responses become conditioned.

- **trait group**
  a highly localised aggregation of interacting conspecific organisms that may differ in the mean of some trait from other such groups and which remains aggregated only temporarily (often for less than a generation).

- **trans**
  of two genetic elements, located on different homologous chromosomes within an individual genome; of gene action, influence of one genetic element on the function or expression of another genetic element located on another chromosome.

- **transcription**
  1. the transfer of information from a DNA molecule into an RNA molecule.
  2. (unit) a unit comprising the entire stretch of DNA that is transcribed into a single RNA molecule marked by the initiation and termination sites of a gene.
  3. in genetics, the conversion of the genetic information in a DNA sequence into a messenger RNA molecule.

- **transduction**
  1. the transfer of genetic material from one cell to another by means of a viral vector.
  2. in the context of sensory systems, the transformation

of the energy of a stimulus to a change in membrane potential of a sensory cell.
3. viral transfer of DNA from one bacterial cell to another; in generalised transduction, recombination occurs between genetic material from two cells without selective transfer, whereas in specialized transduction, selective transfer occurs from DNA located near the integration site of the prophase.

- **transfection**
infection of a cell with isolated DNA or RNA from a virus or viral vector.

- **transfer efficiency**
efficiency with which energy is passed through various steps in the trophic structure of a community.

- **transfer function**
variation in the output level on a signal transmitter after a defined change in the input.

- **transfer RNA (tRNA)**
an RNA molecule that functions as an interpreter between nucleic acid and protein language by picking up specific amino acids and recognising the appropriate codons in the mRNA.

- **transformation**
1. the introduction of an exogenous DNA preparation (transforming agent) into a cell.
2. The conversion of a normal animal cell to a cancerous cell.
3. (bacterial) A phenomenon in which external genetic material is assimilated by a cell.

- **transgenic**
1. an organism containing a gene which has been artificially transferred from a member of another species.
2. an organism containing certain genes from another species; produced, for example, by injecting foreign DNA into the nuclei of egg cells or early embryos.

- **transgression**
the rising of sea levels to cover continental land areas thus greatly increasing the area of inshore communities. See **regression**.

- **transient polymorphism**
the occurrence of two (or more) forms of a species or of genes (alleles) within a population while one form

is being replaced by another.

- **transition probability**
  a measure of the likelihood p that a given behaviour pattern will occur given that another behaviour pattern q will follow.

- **translation**
  1. the process in which the genetic code contained in the nucleotide sequences of mRNA directs the order of amino acids in the formation of peptides.
  2. the transfer of information from an RNA molecule into a polypeptide, involving a change of language from nucleic acids to amino acids.

- **translocation**
  1. exchange of parts of chromosomes among non-homologous chromosomes.
  2. an aberration in chromosome structure due to a mutagen or an error in meiosis-a chromosomal fragment attaches to a nonhomologous chromosome.
  3. when the RNA carrying the polypeptide moves from the A site to the P site on the ribosome during protein synthesis.
  4. the transport of food via phloem.

- **transmissible disease**
  a disease that is caused by living organisms (such as bacteria, viruses, and parasitic worms) and that can spread from one person to another by air, water, food, body fluids (or in some cases by insects or other organisms).

- **transmission electron microscope**
  an instrument which uses a focussed electron beam to examine ultra structural features of thin sections of cells.

- **transmission threshold**
  the basic reproductive rate of a parasite that is a necessary condition for a disease to spread.

- **transpecific evolution**
  evolutionary phenomena and processes beyond the species level, such as the origin of new higher taxa, new organs, evolutionary trends, the extinction of faunas and the like.

- **transpiration**
  the evaporation of water from a plant surface.

- **transport vesicle**
  a vesicle that moves material from the ER to the Golgi apparatus.

- **transposable element**
  a segment or fragment of DNA that can move from one position in the genome to another.

- **transposan**
  a transposable element.

- **transposition**
  movement from one site in the genome to another.

- **transposon**
  a transposable genetic element; a mobile segment of DNA that serves as an agent of genetic change.

- **transverse orientation**
  orientation at a temporarily fixed angle to the direction of the external stimulus or at a fixed angle.

- **transversion**
  a mutation caused by the substitution of a pyrimidine for a purine or vice versa.

- **traplining**
  a feeding strategy during which an animal visits a sequence of widely dispersed food sources, obtaining a small part of its daily food requirement at each feeding location.

- **treading**
  milk elicitation movement in many mammals. They use either their front or hind feet to paw at the teat during nursing. In chickens behaviour of males displayed in the copulatory mounting position prior to contact, characterised by several relatively fast movements of the legs.

- **treatment area**
  the site- specific location of a resource improvement activity.

- **treatment effect**
  in statistics, the change in response produced by the particular variable that was isolated by an experimental manipulation.

- **tree opening**
  an opening in the forest created by even-aged silvicultural practices.

- **trench**
  the trough at the zone of subduction.

- **trial**
  a single, experimentally manipulated occasion on

which an instance of a specified response is elicited or may occur.

- **trial and error learning**
behavioural plasticity. See **operant conditioning**

- **trial-and-error learning**
1. operant conditioning as it is observed to occur under conditions where relatively precise experimental controls have not been established.
2. a mode of learning by repeated experience.

- **triassic**
1. a geological era lasting from approximately 250 to 213 million years ago.
2. a geological period of the Mesozoic era 240-210 my ago.

- **trichome**
single- or many-celled outgrowth from the epidermis of a plant. A plant hair.

- **triplet code**
a set of three-nucleotide-long words that specify the amino acids for polypeptide chains.

- **triploblastic**
possessing three germ layers: the endoderm, mesoderm, and ectoderm. Most eumetazoa are triploblastic.

- **triploid**
having three times the number of chromosomes in the haploid complement.

- **trisomic**
a type of aneuploidy in which one chromosome is present in triplicate in the fertilised egg.

- **triumph ceremony**
epigamic behaviour pattern during pair formation in geese. The male at first makes sham attacks toward objects that are normally avoided. Following such an attack he triumphantly returns to his intended mate and threatens beyond her. If she joins the triumph ceremony, a defensive alliance has been formed which is a part of pairing. After pairing this behaviour continues to maintain the pair bond but functions as a greeting ceremony.

- **troglodyte**
a subterranean organism, cave dweller.

- **trophallaxis**
mouth to mouth exchange of food in social insects.

• **trophic**
1. (level) the division of species in an ecosystem on the basis of their main nutritional source.
2. (structure) the different feeding relationships in an ecosystem that determine the route of energy flow and the pattern of chemical cycling.

• **trophic level**
position in the food chain assessed by the number of energy-transfer steps to reach that level.

• **trophic structure**
the organisation of a community described in terms of energy flow through its various trophic levels.

• **trophobiosis**
the relationship in which ants receive honeydew from aphids and other homopterans or the caterpillars of certain butterflies.

• **trophoblast**
the outer epithelium of the blastocyst, which forms the fetal part of the placenta.

• **tropic hormone**
a hormone that has another endocrine gland as a target.

• **tropical dry forest**
a type of forest found near the Equator that has distinct rainy and dry seasons. Many tropical dry forest plants are adapted to withstand high temperatures and seasonal droughts.

• **tropical rain forest**
1. a type of wet forest found near the Equator that harbours the richest diversity of terrestrial plant and animal species.
2. the most diverse of all communities, located nearer the equator where rainfall is abundant; harbours more species of plants and animals than all other terrestrial biomes combined.

• **tropism**
1. the spatial orientation of sessile organisms due to differential growth.
2. a growth response that results in the curvature of whole plant organs toward or away from stimuli due to differential rates of cell elongation.

• **tropotaxis**
a type of orientation. The process by which animals make a simultaneous com-

parison of the intensity of stimulation received from two or more receptors and can strike a balance between them. Two organs are differently stimulated, the difference causing a reaction bringing the animal in line with the stimulus where the organs are equally stimulated.

- **true predator**
  a predator which kills other organisms (their prey) more or less immediately after attacking them, killing several or many over the course of its lifetime.
- **TSI (Timber Stand Improvement)**
  actions to improve growing conditions for trees in a stand, such as thinning, pruning, prescribed fire or release cutting.
- **tumour**
  a mass that forms within otherwise normal tissue, caused by the uncontrolled growth of a transformed cell.
- **tundra**
  the biome that occurs around the Arctic circle, characterised by lichens, mosses, sedges and dwarf trees.
- **tundra**
  a biome at the extreme limits of plant growth; at the northernmost limits, it is called arctic tundra, and at high altitudes, where plant forms are limited to low shrubby or mat like vegetation, it is called alpine tundra.
- **tunicate**
  a member of the Subphylum Urochordata (Phylum Chordata), comprising sea squirts and similar forms.
- **turbulence**
  fluid flow in which the motion at any point varies rapidly in direction and magnitude.
- **turgid**
  firm; walled cells become turgid as a result of the entry of water from a hypo-osmotic environment.
- **turgor**
  the distension of living tissue due to internal pressures.
- **turgor pressure**
  force directed against a cell wall after the influx of wa-

ter and the swelling of a walled cell due to osmosis.

- **type conversion**
  the conversion of the dominant vegetation in an area from forested to non-forested or from one species to another.

- **type I error**
  rejecting the null hypothesis when it is in fact true (concluding there is an effect when none exists).

- **type II error**
  accepting the null hypothesis when it is false (i.e. failing to detect a real effect).

- **typical intensity**
  some displays are performed in a very stereotyped manner despite the fact that the underlying motivation varies. These displays are said to be released at a typical intensity.

- **typological**
  a type of thinking based on a concept in which variation is disregarded.

- **typology**
  a method of classification or study of natural groups based on the assumption that all members of a taxonomic unit conform to a given morphological plan without significant variation.

- **ultimate causation**
  the hypothetical evolutionary explanation for the existence of a certain pattern of animal behaviour.

- **ultimate determinants of behaviour**
  the evolutionary factors (adaptiveness and phylogenetic origins) that act to maintain the behaviour in the population. See **Proximate Determinants.**

- **ultrasound**
  sounds at frequencies above the range of human hearing.

- **ultraviolet light**
  the wavelengths of light that are just below the range of human vision but can be seen by bees and many other arthropods.

- **umwelt**
  1. refers to the perceptual world of animals. Used in the context that an animal's umwelt is different from that of a human. The Umwelt is made up of the *Merkwelt*, the different sensory and processing systems and its

Wirkwelt or effector world, the different ways of responding to the environment.

2. sensory world. Subjective set of stimuli to which an animal is responsive in a given motivational state. Each species has evolved responses only to those stimuli that prove relevant. It is this simpler world that actually falls within the animal's perception at any particular moment.

- **unconditioned response**
  a regular and measurable response elicited by the unconditioned stimulus in the classical conditioning experiment.

- **unconditioned stimulus**
  in classical conditioning, a stimulus which evokes or elicits a regular and measurable response (the unconditioned response.)

- **underburn**
  a burn by a surface fire that can consume ground vegetation and ladder fuels.

- **undercompensating density dependence**
  density dependence in which death rate increases, or birth or growth rate decreases, less than initial density increases, initial density still lead to (smaller) increases in final density.

- **understory**
  the trees and woody shrubs growing beneath the overstorey stand of trees.

- **uneven-aged management**
  actions that maintain a forest or stand of trees composed of intermingling trees that differ markedly in age. Cutting methods that develop and maintain uneven-aged stands are single-tree selection and group selection.

- **uniformitarianism**
  a theory of geology based on the premise that the geological processes that built past formations do not differ from those operating today (such as erosion and deposition).

- **uniformity**
  the principle that laws of nature in operation at present are the same as those which have been operational throughout time.

- **unimodal**
  a frequency distribution with one peak.

- **uniparous**
  producing one offspring at a time.

- **unit leaf rate**
  the dry weight increment made by a plant over a period of time expressed as a function of the plant's leaf area.

- **unitary organisms**
  those that proceed by a determinate pathway of development of a tightly canalised adult form, e.g. all arthropods and vertebrates. The contrast is with 'Modular organisms' in which growth occurs by the indeterminate iteration of repeated units of structure (modules).

- **univoltine**
  having only one generation per year.

- **unlearned behaviour**
  1. species-specific behaviour, the necessary and sufficient antecedents of which are unknown to, and often of little interest to, the investigator.
  2. behaviour that has been experimentally demonstrated, in both its stimulus control and topography, to be independent of and unmodified by the operation of variables encountered in conditioning and learning, such as the occurrence of reinforcing stimuli.

- **unregulated harvest**
  tree harvest that is not part of the allowable sale quantity. It can include the removal of cull or dead material or noncommercial species. It also includes volume removed from non-suitable areas for research, to meet objectives other than timber production (such as wildlife habitat improvement), or to improve administrative sites (such as campgrounds.).

- **unsaturated fatty acid**
  a fatty acid possessing one or more double bonds between the carbons in the hydrocarbon tail. Such bonding reduces the number of hydrogen atoms attached to the carbon skeleton.

- **unstable equilibrium**
  1. in an ecological context, a level of a population, or

populations, or of resources, from which slight displacements lead to larger displacements.
2. a frequency equilibrium between alleles in a population that is readily upset because of a selection disadvantage of the heterozygotes.

- **unsuitable lands**
forest land that is not managed for timber production. Reasons may be matters of policy, ecology, technology, silviculture or economics.

- **upwelling**
an upward flow of cold, heavy deep-sea water caused as offshore currents draw away warm surface water. The deep-sea water of upwellings is generally rich in nutrients.

- **urban heat island**
build up of heat in the atmosphere above an urban area. This is produced by the large concentration of cars, buildings, factories and other heat-producing activities.

- **urea**
a soluble form of nitrogenous waste excreted by mammals and most adult amphibians.

- **ureter**
a duct leading from the kidney to the urinary bladder.

- **urethra**
a tube that releases urine from the body near the vagina in females or through the penis in males; also serves in males as the exit tube for the reproductive system.

- **uric acid**
an insoluble precipitate of nitrogenous waste excreted by land snails, insects, birds and some reptiles.

- **urine marking**
a widespread behaviour pattern among mammals in which the paths and boundaries of the territory are marked with urine.

- **urochordate**
a chordate without a backbone, commonly called a tunicate, or sea squirt, a sessile marine animal.

- **use value**
value derived from actual use of a good or service. Uses may include indirect uses. For example, enjoying a television show about whales provides an indirect use value for the whales.

- **use, allowable**
  an estimate of proper range use. Forty to fifty percent of the annual growth is often used as a rule of thumb on ranges in good to excellent condition. It can also mean the amount of forage planned to be used to accelerate range rehabilitation.

- **uterus**
  a female reproductive organ where eggs are fertilised and/or development of the young occurs.

- **utility**
  the animal's evaluation of the attractiveness of the alternative choices.

- **vaccination**
  the production of active immunity by administration of a vaccine.

- **vaccine**
  a harmless variant or derivative of a pathogen that stimulates a host's immune system to mount defences against the pathogen so as to confer immunisation.

- **vacuole**
  1. membrane-bound, fluid-filled sac within the cytoplasm of a cell.
  2. a membrane-enclosed sac taking up most of the interior of a mature plant cell and containing a variety of substances important in plant reproduction, growth, and development.
  3. a liquid filled cavity in a cell, enclosed within a single membrane. Various types exist including those used for storage, intracellular digestion and osmoregulation.

- **vacuum activity**
  1. when an instinctive behaviour has not occurred for an extended period of time, the threshold value for relating stimuli is lowered. This reduction in threshold may reach a point at which the response occurs without a demonstrable stimulus and hence the term vacuum activity.
  2. the occurrence of a fixed action pattern in the apparent absence of its usual releaser or sign stimulus.

- **vagility**
  the dispersal ability of a species.

- **vagina**
  a thin-walled chamber that forms the birth canal and is

the repository for sperm during copulation.

- **vagrancy**
  wide dispersal where average propagule moves more than 30 homeranges away from natal site.

- **valence shell**
  the outermost energy shell of an atom, containing the valence electrons involved in chemical reactions of that atom.

- **variability, coefficient of**
  in statistics the standard deviation expressed as a percentage of the mean.

- **variable costs**
  production costs that change when the level of production changes, so that when more is produced the costs increase; as opposed to fixed costs.

- **variance**
  the average squared deviation of an observation from the arithmetic mean; hence, a measure of variation. $S^2 = (\text{mean} - x)^2 N-1$.
  where N is the number of observations.

- **variance enhancement**
  increase in the variance of the reproductive potential among members of a parent's brood. The implication is that more brood members are thereby brought to lie above or below the threshold for breeding, thus increasing the proportion of the brood becoming breeders or potential helpers respectively.

- **variance utilisation**
  utilisation by parents of aid from those of their offspring that are unable to breed for reasons not imposed by the parents. The implication is that offspring vary in their reproductive potential, some being above the threshold for attempting breeding and some below.

- **variety class**
  a way to classify landscapes according to their visual features. This system is based on the premise that landscapes with the greatest variety or diversity have the greatest potential for scenic value.

- **variety**
  1. in Zoology an ambiguous term for a set of phenotypic variants, whether they be nongenetic variants, mor-

phs, domestic breeds or races.
2. in Botany used to denote a variety in the zoological sense or, more commonly, a subspecies.

- **variolation**

a procedure practiced by the ancient Chinese in which humans are exposed to scabs of smallpox (also called variola) survivors, thus conferring immunity in about 99% of the attempts.

- **vas deferens**

the tube in the male reproductive system in which sperm travel from the epididymis to the urethra.

- **vascular**

1. (cambium) a cylinder of meristematic cells surrounding the xylem that produces secondary xylem and phloem.
2. (plants) plants with vascular tissue, (all modern species except the mosses and relatives.)
3. (tissue) plant tissue consisting of cells joined into tubes which transport water and nutrients.

- **vector**

1. any agent (living or otherwise) that acts as a carrier for a pathogenic organism and transmits it to a susceptible host.
2. a physical quantity with a direction as well as a strength.
3. an agent consisting of a DNA molecule known to autonomously replicate in a cell to which another DNA segment may be attached experimentally so as to bring about the replication of the attached segment.
4. a living organism which actively transmits another organism to a host;
5. a cloning vector is an agent used to transfer DNA in genetic engineering, such as a plasmid that moves recombinant DNA from a test tube back into a cell or a virus that transfers recombinant DNA by infection.

- **vegetal pole**

the end of a zygote opposite the animal pole; the yolky end.

- **vegetation management**

activities designed primarily to promote the health of

forest vegetation for multiple-use purposes.

- **vegetation type**
  a plant community with distinguishable characteristics.

- **vegetative**
  a type of reproduction involving cloning of plants by asexual means.

- **veins**
  vessels that return blood to the heart.

- **veld**
  open pasture land in South Africa.

- **Venn diagram**
  a diagram that makes use of set theory to organize relationships; usually portrayed as three intersecting circles.

- **ventilation**
  any method of increasing contact between the respiratory medium and the respiratory surface.

- **ventral**
  the bottom (or belly) half of a bilaterally symmetric animal.

- **ventral earth reaction**
  orientation so that gravitational force acts perpendicularly to long and transverse axes of the body. The dorsal surface is usually kept uppermost.

- **ventricle**
  1. a chamber that pumps blood out of the vertebrate heart.
  2. a space in the vertebrate brain filled with cerebrospinal fluid.
  3. one chamber among the several fluid-filled compartments of the vertebrate brain.

- **venules**
  vessels into which capillaries converge and that empty into veins.

- **vertebrates**
  chordate animals with backbones, consisting of the mammals, birds, reptiles, amphibians and various groups of fishes.

- **vertical diversity**
  the diversity in a stand that results from the different layers or tiers of vegetation.

- **Vesicular Arbuscular Mycorrhiza (VAM)**
  an intimate and perhaps usually mutualistic association between a fungus and a plant root in which the

fungus enters the host cells and also usually extends widely into the surrounding soil.

- **vessel elements**
specialised short, wide cells in angiosperms, arranged end to end to form continuous tubes for water transport.

- **vestigial organ**
a type of homologous structure that is rudimentary and of marginal or no use to the organism.

- **vestigial structures**
morphological traits that were once functional in the phylogenetic history of the organism but now serve no useful function to the organism.

- **viability**
a relative measure of the number of surviving individuals of any given phenotypic or genotypic class.

- **viable**
the capability of living or continuing development.

- **viable population**
the number of individuals of a species sufficient to ensure the long-term existence of the species in natural, self-sustaining populations that are adequately distributed throughout their range.

- **vicariance**
species evolved by allopatric speciation.

- **vicariance biogeography**
a method for assisting the degree of congruence in the distributional patterns of taxa. This method reveals common elements in distributional patterns and is used to reconstruct the evolutionary history of a taxon.

- **vicariance event**
the splitting or division of a species through the development of a natural biogeographical barrier.

- **vigilance**
conspicuous behaviour associated with watching out for predators. Refers to an animal's state of readiness to detect certain specified events occurring unpredictably in the environment.

- **virchow**
contributed to the formulation of the Cell Doctrine.

- **virgin**
an animal that has not copulated in its lifetime. Primarily used for females.

- **virgin forest**
a natural forest virtually uninfluenced by human activity.

- **virion**
a complete virus particle that represents the extracellular phase of the virus life cycle.

- **viroid**
a plant pathogen composed of molecules of naked RNA only several hundred nucleotides long.

- **virulence**
the capacity of a pathogen to invade host tissue and reproduce; the degree of pathogenicity.

- **virulent**
a virus capable of causing a disease state.

- **virus**
a nonliving particulate infective agent smaller than a bacterium and incapable of independent reproduction. It is replicated only within host cells. Viruses are important causative agents of diseases in all types of living organisms.

- **visceral muscle**
smooth muscle found in the walls of the digestive tract, bladder, arteries and other internal organs.

- **visible light**
that portion of the electromagnetic spectrum detected as various colours by the human eye, ranging in wavelength from about 400 nm to about 700 nm.

- **vision**
the detection of light by eyes and the behavioural responses it produces.

- **visual cliffs**
various nidifugous animals show an innate fear of a precipice which they recognize from the visual impression alone without themselves facing an experience with falling.

- **visual quality objective**
a set of measurable goals for the management of forest visual resources.

- **visual resource**
a part of the landscape important for its scenic quality. It may include a compos-

ite of terrain, geologic features or vegetation.

- **vital attributes (species)**
  properties of species that determine their place in a succession.

- **vitalism**
  1. the view that phenomena exhibited by organisms are the result of supernatural forces as distinct from chemical and physical forces that exert directional effects on variation and evolution.
  2. the belief that natural phenomena are governed by a life force outside the realm of physical and chemical laws.

- **vitamin**
  an organic molecule required in the diet in very small amounts; vitamins serve primarily as coenzymes or parts of coenzymes.

- **viviparous**
  referring to a type of development in which the young are born alive after having been nourished in the uterus by blood from the placenta.

- **vocal mimicry**
  a term used in the bioacoustical literature for a species' habit of adding to its vocal repertoire sounds imitated from another species. According to the theory, Beau Geste hypothesis, vocal mimicry may give a false indication of higher population density to both conspecifics and members of other species in competition with the singer.

- **vocal repertoire**
  sound repertoire. Inventory of the sound patterns produced by an individual or a species.

- **vocal simulation**
  voice simulation; sound simulation. The playback of natural or artificial sounds, as a rule by tape recorder, to test their effects on the behaviour of animals being studied for auditory responsiveness. By manipulating the sounds (transposing components in sound sequences, changing rhythm, filtering out certain frequencies), one can determine, from the reactions of the test animals, which are charac-

teristic of a naturally occurring sound are significant to its effects and which are not.

- **vocalisation**
production of songs, calls and other sounds by animals. The production of sound by means of a vocal apparatus.

- **volant**
adapted for flying or gliding.

- **voltage**
electrical potential able to move charge in a current; Unit: volt.

- **voltage-gated channel**
ion channel that opens or closes according to the level of the voltage difference across the neuronal membrane.

- **voltage-sensitive channel**
an ion channel in the membrane of excitable cells that is constructed of protein molecules and allows only certain ions to pass under specific changes of electric potential; also called voltage-gated channel.

- **voltine**
pertaining to the number of broods produced over the year. Bivoltine refers to two broods; univoltine to one.

- **voltinism**
a polymorphism in a species in which some members of the population enter diapause whereas others do not.

- **vomeronasal organ**
just above the hard palate, the nasal mucosa covers this secondary organ of olfaction. The behaviour called flehmen helps odours get into the incisive ducts of the vmo by opening the mouth and partially obstructing the nostrils.

- **von baer's law**
development proceeds from the general to the special. The earliest embryonic stages of related organisms are identical; distinguishing features are added later as heterogeneity differentiates from homogeneity. Recapitulation is impossible; young embryos are undifferentiated general forms, not previous adult ancestors.

- **v_x**
reproductive value; of an individual of a specific age, its likely contribution to the growth of the population.

- **w**
 the fitness of a given genotype; the relative rate by which the frequency of a given genotype is increased each generation by selection.

- **waggle dance**
 refers to the communication behaviour of bees indicating the direction and distance to a food source which is more than 100 m away.

- **wallowing**
 behaviour in which an animal spends time rolling and lying in mud or muddy water.

- **war of attrition**
 a game in which winners are decided simply by a contest involving waiting, or displaying, for different lengths of time. The solution is a mixed ESS of variable waiting times.

- **warm-up**
 over the first few occasions within a short period of time that a specified response occurs, response-strength may increase relatively rapidly, independently of reinforcement contingencies. Such increments in response-strength are termed warm-up.

- **warning behaviour**
 alarm behaviour. Behaviour performed in response to predators.

- **water potential**
 in osmosis, the tendency for a region to take up water from pure water, through a selectively semipermeable membrane. Water flows toward the region with the more negative water potential.

- **water table**
 the upper surface of groundwater. Below it, the soil is saturated with water.

- **water yield**
 the runoff from a watershed, including groundwater outflow.

- **watershed**
 the entire region drained by a waterway (or into a lake or reservoir. More specifically, a watershed is an area of land above a given point on a stream that contributes water to the stream flow at that point.

- **water-vascular system**
 the system of canals and tube like appendages that

serves as the circulatory system, locomotory system and ingestion system of many echinoderms. An opening to the system is connected with the surrounding sea water.

- **wavelength**

  1. the horizontal distance between the crests of adjacent waves. Wavelength of light refers to colour and wavelength of sound refers to pitch.
  2. the distance between successive peaks of a wave train, such as light or another type of electromagnetic radiation. Usually designated lambda, l, and often measured in nanometers, nm.

- **w-chromosome**

  the sex chromosome which in the case of female heterogamety is present in the female only.

- **weaning**

  the break of an offspring's dependence on its parents.

- **weed species**

  plants that threaten human welfare by competing with other plants that have food, timber or amenity value.

- **weighting**

  a method of attaching different importance values to different characters.

- **wetlands**

  areas that are permanently wet or are intermittently covered with water.

- **wetlands**

  areas that, at least periodically, have waterlogged soils or are covered with a relatively shallow layer of water. Bogs, freshwater and saltwater marshes, and freshwater and saltwater swamps are examples of wetlands.

- **wheel posture**

  copulation posture in dragonflies which is determined by the unique position of the male copulatory organ on the second abdominal sternite.

- **whitten effect**

  synchronisation of the oestrus cycling of females housed together.

- **wild type**

  1. the allele, genotype, or phenotypes belonging to the same population.
  2. an individual with the normal phenotype.

- **wilderness (wilderness area)**
  undeveloped federal land retaining its primeval character, without permanent human habitation or improvements. It is protected and managed to preserve its natural condition.

- **wildfire**
  any wildland fire that is not a prescribed fire.

- **wildlife habitat diversity**
  the distribution and abundance of different plant and animal communities and species within a specific area.

- **willingness to pay**
  the amount—measured in goods, services or dollars—that a person is willing to give up to get a particular good or service.

- **windthrow**
  trees uprooted by wind.

- **wiping reflex**
  if frogs or toads eat some oversized or unpalatable object, they try to expel it by opening the mouth wide and moving the forelegs forward from the side in a wiping movement.

- **wobble**
  a violation of the base-pairing rules in that the third nucleotide (5' end) of a tRNA anticodon can form hydrogen bonds with more than one kind of base in the third position (3' end) of a codon.

- **wood fibre production**
  the growing, tending, harvesting and regeneration of harvestable trees.

- **woodland products**
  harvestable items from pinyon-juniper woodlands. These include fuelwood, posts, pine nuts and Christmas trees.

- **work**
  a process which requires the transformation of energy.

- **worker**
  among social insects an individual that does not or rarely reproduces and instead helps another individual (usually the mother) to rear offspring, construct nests, defend sites etc.

- **working hypothesis**
  a hypothesis that serves as the basis for future experiments or observations.

- **xanthophyll**
  a yellow or orange pigment commonly found as an accessory photosynthetic pigment.

- **x-chromosome**
  the sex chromosome which in cases of male heterogamety is present in both sexes.

- **xenobiosis**
  the relation in which colonies of one ant species live in the nests of another and move freely among the hosts obtaining food from them by trophallaxis or other means.

- **xenoparasite**
  a parasite infesting an organism that is not its normal host.

- **xeric**
  1. having little or no moisture. tolerating or adapted to dry conditions.
  2. a habitat in which water is in very limited supply. Organisms dwelling in such habitats exhibit adaptations which tend to limit desiccation.

- **xerophyte**
  a plant adapted to an environment with a limited water supply.

- **yarding**
  moving the cut trees from where they fell to a centralized place (landing) for hauling away from the stand.

- **yawning**
  wide opening of the mouth thought to be associated with a specific physiological function.

- **y-chromosome**
  the sex chromosome which in cases of male heterogamety is present in the male only.

- **yeast**
  a unicellular fungus that lives in liquid or moist habitats, primarily reproducing asexually by simple cell division or by budding of a parent cell.

- **yolk**
  1. the stored food in animal eggs, usually rich in protein and lipid.
  2. (sac) one of four extraembryonic membranes that supports embryonic development; the first site of blood cells and circulatory system function.

- **z-chromosome**
  the sex chromosome which in cases of female heterogamety is present in both sexes.

- **z-DNA**
  a form of DNA in which the molecule spirals to the left rather than to the right.

- **zeitgeber**
  time giver. An external stimulus that brings an animal's endogenous periodicity into synchrony with environmental periodicity.

- **zenith**
  the point on the celestial sphere directly above the observer or above a given point on the surface.

- **Zero Net Growth Isocline (ZNGI)**
  an isocline along which the rate of population growth is zero.

- **Zero Population Growth (ZPG)**
  state in which the birth rate (plus immigration) equals the death rate (plus emigration) so that the population of a geographical area is no longer increasing.

- **zig zag dance**
  the courtship behaviour of the male stickleback in which he alternately moves toward the nest or female and then away.

- **zirkeln (prying movements)**
  a feeding movement for birds in which the bird inserts its beak in a crevice in the ground, e.g., between grass roots or under the bark of trees, and then forcefully opens the beak thus expanding the hole and increasing its access to worms and insects. Conspicuous in starlings.

- **zoarium**
  a colony of bryozoans.

- **ZOI (Zone Of Influence)**
  the area influenced by Forest Service management activities.

- **zonation**
  the characteristic distributions of species along environmental gradients.

- **zone of intergradation**
  the region of contact between two populations of contrasting phenotypes, in which a mixture of phenotypes occurs through interbreeding; sometimes called zone of hybridisation.